Transnational Cinematography Studies

Transnational Cinematography Studies

Edited by
Lindsay Coleman, Daisuke Miyao
and Roberto Schaefer

Preface by
Roberto Schaefer

LEXINGTON BOOKS
Lanham • Boulder • New York • London

Published by Lexington Books
An imprint of The Rowman & Littlefield Publishing Group, Inc.
4501 Forbes Boulevard, Suite 200, Lanham, Maryland 20706
www.rowman.com

Unit A, Whitacre Mews, 26-34 Stannary Street, London SE11 4AB

British Library Cataloguing in Publication Information Available

Library of Congress Cataloging-in-Publication Data

Names: Coleman, Lindsay, 1978- editor. | Miyao, Daisuke editor. | Schaefer, Roberto editor.
Title: Transnational cinematography studies / edited by Lindsay Coleman, Daisuke Miyao, and Roberto Schaefer.
Description: Lanham : Lexington Books, 2016. | Includes bibliographical references and index.
Identifiers: LCCN 2016041507 (print) | LCCN 2016053872 (ebook) | ISBN 9781498524278 (cloth : alk. paper) | ISBN 9781498524285 (Electronic)
Subjects: LCSH: Cinematography | Digital cinematography.
Classification: LCC TR899.T77 2016 (print) | LCC TR899 (ebook) | DDC 777/.8–dc23
LC record available at https://lccn.loc.gov/2016041507

⊚™ The paper used in this publication meets the minimum requirements of American National Standard for Information Sciences—Permanence of Paper for Printed Library Materials, ANSI/NISO Z39.48-1992.

Printed in the United States of America

Contents

Preface

Transnational Cinema—An Introduction and a Question

Roberto Schaefer

The concept of Transnational Cinema is not new. In fact it is almost questionable as to when cinema was *not* transnational. When the art and practice of cinema began, most films were indeed national in the respect that they were made by a small group of close-knit people largely in one location, France for example. As the world grew older by leaps and bounds and wars spread across national borders, there were larger than ever migrations of established people, many in the film business. During World War I there began a movement out of Eastern Europe and Germany mostly toward the West. At that time many cinema professionals were displaced and ended up in Hollywood, the center of the Western world's motion picture business at the time. This migration continued as more opportunities were discovered and a second world war accelerated and intensified the migrations. Some of these professionals ended up in the Far East, and Australasia. But the greater percentage settled in Hollywood. So it is therefore arguable that motion pictures made in the United States of America, predominately in Hollywood, California, were already transnational from the 1920s. Many Hollywood directors, writers, actors and cinematographers as well as composers who came from Germany, Hungary, Poland, and the rest of war-torn Europe helped to shape the whole aesthetic of Hollywood in the 1930s and 1940s and 1950s.

Actors, directors, designers, cinematographers and more moved to where they could live and work relatively freely. There were also migrations of these artists to Mexico and South America, Luis Bunuel being one shining example.

It may be safe to say that possibly only China and Japan held on to their own National Cinemas longer than most other cultures and geographical locations did.

As we entered the mid to late twentieth century, most bets were off. Almost every country had its share of motion picture professionals working or studying in foreign lands. By doing so they transcended the national borders of thought and vision and were able to embrace the new lands by way of cultural influences on story, character, and visuals. This cross pollination had no financial borders either. From the smallest independent 16mm production in NYC to the largest 35mm or even 65mm blockbuster in Morocco, the influences and physical presences were obvious. Look at Amos Poe's 1976 film "Unmade Beds" shot in black and white in NYC pretending to be Paris and you can't help but see the influences of Godard and Truffaut. He also had a pan-national creative team composing, photographing, and editing the film.

Almost every film being made in the world today is transnational. This is partly due to the ease of international travel, internet communication, and information and monetary incentives. This book will look at various issues and influences on specific cinematographers and directors, and how they fit into the transnational cinema concept.

Introduction

Daisuke Miyao

What is the status of cinematography in the digital age? In *Transnational Cinematography Studies*, we tackle with diverse questions that extend from the areas of technology to those of emotions and affect: What will the dominance of CGI digital cinematography bring to the "actual" cinematography? What effect on the visual palette does the governing aesthetic of "fewer lights" have? With increasingly mobile, lightweight cameras, how has the style of cinematography crossed over into new forms of camera acrobatics? What is the status of documentary under such conditions? What is to be made of a documentary such as *Leviathan* (2014), which has a compelling aesthetic, yet one created accidentally, effectively by the technology itself? Can cinematography increasingly withhold information? What are the relationships between such choices of framing, camera movement, and focal length and emotions? How are cinematographers working different systems to share information, technologies, and techniques? What impacts does the transnational media industry have on cinematography?

By answering these questions, in *Transnational Cinematography Studies* we introduce new perspectives to the discipline of film and media studies. First, this volume's focus is titled "Cinematography Studies." It focuses on a crucial yet largely unexplored area in film and studies: the substantial communication between critical studies of cinema and film production practices. Even though books have been published on cinematography, most of them are memoirs/interviews of Hollywood and European cinematographers and production manuals. Despite cinema's innate status as a technological medium, scholars of film and media studies have focused more on analyses of stories and themes, and their relations to sociopolitical and economic contexts. When styles are discussed, they have mostly been analyzed aesthetically (relationship to realism, etc.). For instance, even the definition of the "director of photography" has not been clearly explained by any scholar to this day. Historically, it is not clear when the title of the "director of photography" was first used in the history of motion pictures. According to the *American Cinematographer*, Roy Henry Klaffki, who worked with such filmmakers as Erich von Stroheim and Lois Weber, "held the title of Director of Photography at Metro Pictures in the early 1920s, shooting films as well as supervising the work of other Metro cinematographers and overseeing quality control in the lab."[1] However, the title was rarely used in credits of films as well as in the articles in the *American Cinematographer* until early to mid-1930s. Either "cameraman" or "cinematographer" was the most common word to describe the profession. Arguably, based on the pages of the *American Cinematographer*, it was James Wong Howe who started calling himself and his profession the director of photography in 1934. In an essay in which he explained his cinematographic strategy in *Viva Villa* (1934), Howe emphasized the significance of "co-operation" between two professions, "the Art Director" and "the Director of Photography," in order to build sets and to suit the camera-treatment "more efficiently."[2] In his 1931 essay, "Lighting," which appeared in the ASC's *Cinematographic Annual Volume Two*, he did not use the term "director of photography" but only "cinematographers."[3] What happened between 1931 and 1934? One possible factor behind the switch was technological. The advent of Supersensitive Panchromatic Type Two Motion Picture Negative Film by Eastman Kodak Company was announced on February 5, 1931. With emulsion reacting "faster" to light—75 percent more speed to blue light, 200 percent more to green light, and from 400 to 500 percent more to red light—the Supersensitive Panchromatic Type Two extended the latitude and enriched the gradation between black and white to be much more diverse.[4] After the announcement, the relationship between lighting and film stock became the central topic of discussions among cinematographers. For instance, Hal Hall, the editor of the ASC's *Cinematographic Annual Volume Two*, wrote:

> Too many cinematographers confuse the improved speed with exaggerated contrast—a serious mistake. The aim of the cinematographer should be for natural tonal contrast with an artistic softness—and this softness cannot be attained with flat lighting. Soft lighting should be used—but it should be normally balanced, sacrificing none of the tonal and physical contrasts which the cameraman wishes to preserve.[5]

As such, this book attempts to integrate theories and practices of cinematographic technology.

Secondly, this book is "Transnational Studies." It expands the scope of film and studies into the arena of transnationalism. Cinema is now discussed in terms of globalization of audiovisual cultures, with regard to such issues as Hollywood film studios' so-called "runaway productions" (substantial filmmaking in locations outside of the United States) and multinational coproductions; Hollywood remakes of Asian horror films or Hong Kong martial arts films; and the growing significance of international film festivals. However, this volume proposes that globalization is not in itself new in the history of cinema, and that cinema has always been at the forefront of transnational culture from the beginning of its history. While academic discourses on the transnational natures of cinema have been increasing, many of them are following the historical constellation of hegemony and capital in cinema, namely Hollywood's place as production and distribution center. Beyond such a dichotomized view of center-periphery, this volume will examine the diversity of ways in which technologies of cinematography in hegemonic Hollywood cinema have been reconfigured in local contexts.

Cinematography, indeed, has been transnational from the beginning of the history of cinema. After the first public fee-paying screening of films to Parisian audiences at the Salon Indien du Grand Café on December 28, 1895, the Lumière brothers sent numerous cinematographers all over the world, traveling with the compact Cinématographe Lumière, and collected a repertoire of films for screenings in France. Collected images included not only such historic sites and picturesque landscapes as the pyramids in Egypt and Niagara Falls in Canada, but also objects and people of French colonies. There were films about military training in Algeria and Indochina, for instance. Newspapers of the period tended to cover the travels of the Lumière cinematographers as an imperial conquest and marveled at how "the entire world" might soon be "the conquest of the Cinématographe Lumière."[6] Apparently, Lumière films were clearly a part of French imperialism that was observed in the fields of geography, archeology, and anthropology in the name of civilizing the world. But among those cinematographers, there were some who moved beyond such a "national" project of France and became transnational figures.

Gabriel Veyre (1871–1936) was one such cinematographer. Veyre was a twenty-five-year-old pharmacist in 1896 when the Lumière brothers recruited him. "His knowledge of chemistry, his interest in electricity and his passion for photography made him the ideal candidate" for the Lumière brothers' global operation, according to the wall label of the exhibition, *Lumière!: Le Cinéma Inventé*, that commemorated the 120th anniversary of the "birth of cinema" (The Grand Palais, Paris, March 27–June 14, 2015).[7] His travels first took him to Latin America: Mexico, Cuba, Colombia, Panama, and Venezuela where he fell ill in Caracas. After a short rest in Lyon, he was sent to the Far East in July 1898. Traveling by way of Canada, he arrived in Japan, and then went on to China and Indochina. Upon his return to France, he learned that the Sultan of Morocco, Moulay Abd El-Aziz, was looking for someone to teach him photography and cinema. He decided to take the position and arrived at the Marrakech palace in March 1901. The assignment was supposed to last six months, but he remained in Morocco for the rest of his life.

It was true that Veyre had Orientalist ideas originated in French colonial thoughts. For instance, Veyre wrote about Japanese women: "In order to fully understand the social status of Japanese women, we need to completely wipe out the sense of Western morality from our minds. . . . Japanese girls are extremely attractive because they combine decency and indecency, obedience and obscenity."[8] The Japanese women of Veyre's conception were nothing but Madame Chrysanthemum whom Pierre Loti imagined in 1887 and Cio-Cio-San in *Madama Butterfly*, the opera that Giacomo Puccini would create in 1904. Veyre also had a clearly Orientalist image of Japan, in which an imaginary Japan had remained ahistorical and timeless; Japan's growing role as an international trading partner dealing directly with Europe and the United States was not a part of his vision. Veyre lamented:

> Whenever I go through a city, I see too many vulgar derby hats. They abandon the beautifully unique hairstyle that they have learned from their ancestors and are imitating Western hats. There are policemen wearing Western-style uniforms that do not fit. There still are delicate and beautiful wooden houses, but many of them are on the verge of being replaced by unstylish and gigantic concrete buildings.[9]

In fact, Veyre took part in the French colonialist project. In Hanoi, which he visited after Japan and China, the Governor General Paul Doumer commissioned him to produce a cinematographic documentary of French Indochina to be shown at the 1900 Paris World Fair.

But at the same time, Veyre was not only a distant observer of different people and their customs. He was willing to transform himself into Mexican, Japanese, Chinese, and eventually Moroccan personas. He learned different languages. He put on different clothes. He tried to eat Japanese food even though he did not like it very much. He published a book, *Dans l'intimité du sultan* (In the Privacy of Sultan) in 1905, in which he depicted the daily life in Morocco in detail, including the Moroccan manners and the customs at the palace. In 1934, he journeyed through Morocco with a camera, the Lumière Autochrome, and depicted hundreds of landscapes and many scenes of everyday life. He also photographed a self-portrait as a Moroccan. Veyre thus became a transnational cinematographer and a global citizen.

All nine chapters of *Transnational Cinematography Studies* are such attempts of integrating theories and practices as well as of locating cinematography in transnational contexts. Even though some authors do not use the term "transnational" in an explicit manner, their arguments are not confined to a "national cinema" but connected to the transnational network of styles, techniques, and technology of cinematography.

The first three chapters explore the relationship between the technological practices and the styles/aesthetics of cinematography. Lindsay Coleman closely examines the text and the production of Lucien Castaing-Taylor and Verena Paravel's *Leviathan* (2014) and argues how the cinematography of this film approaches the "new documentary aesthetics." For Coleman, *Leviathan* is a documentary yet, remarkably, it is more heralded as an aesthetic experience, a breakthrough in cinematic sensations. Remarkably this film was not "shot" in a conventional manner. There was no specific operator or cinematographer. Instead cameras were installed on the bodies of the film's subjects. Yet by design or not, this approach went some way to produce a distinct, layered aesthetic, one further enhanced by the presence of odd sound artifact generated by the camera, which in turn complement unusual camera movements or transitions from one setting to another. This chapter explores the degree to which such a "laissez-faire" approach does indeed contribute to a defined aesthetic, and also attempts to answer, in cogent terms, the question as to whether it is the camera itself, rather than any particular authorial intervention, which constructs this unique aesthetic.

Jakob Isak Nielsen focuses on one technique of cinematography: camera movement. Nielsen claims that camera movements—real or simulated—are almost as old as the history of film itself. Today camera movements permeate almost any art form that involves moving images: not only films but also TV genres, computer games, music videos, commercials, online videos, and so forth. In his chapter, Nielsen proposes that there are five main functions in camera movement. He elaborates these five functions in the historical context of transnational cinematography, with reference to concrete examples. While the chief focus is on narrative cinema, Nielsen also argues that many of the proposed functions have wider applicability.

My own chapter turns to 3-D cinematography. The history of cinematography has been all about how to capture the three-dimensional world within the two-dimensional frame. Today 3-D appears to be the dominant mode of filmmaking and exhibition practices. Examining the history of cinematography and its relationship to painting and photography, I demonstrate the awareness in 3-D is nothing new in the history of Japanese cinema in its transnational context. Discussing the cinematography of depth from a historical perspective, this essay introduces one aspect of such complications in visual art and technology in Japan. Like Murakami Takashi's superflat theory, my focus is also on the movement of the eyes. However, I do not argue that the movement is limited to the surface no matter how flat the images of Japanese art and cinema look compared to those in Western counterparts. Even in those flat-looking images, cinematography invites the eyes of the viewers to become more unstable, mobile, and ephemeral not only on the surface level but also on the depth of fields.

The authors of the following three chapters discuss the collaborative roles of cinematographers in the practices of filmmaking. For Evan Lieberman, from the first decades of cinema, the innovation and advancement of what might be termed the cinematic language has been accomplished not solely by directors, as is so often suggested by film historians, but rather by directors and directors of photography working in tandem, often in creative relationships that lasted several decades. The collaboration between director D.W. Griffith and cameraman Billy Bitzer may be the first and most notable of such pairings, precipitating a veritable revolution in the techniques of cinematic representation, but also integral to this early movement are Charlie Chaplin and Rollie Totheroh, Sergei Eisenstein and Eduard Tisse, Buster Keaton and Elgin Lessley, F.W. Murnau and Karl Freund, and Jean Vigo and Boris Kauffman. This trend of innovative director/director of photography pairings continues into the contemporary era, but perhaps more significant at this point than the advancement of the visual techniques of cinema, is the way in which the current director/director of photography pairings represent the movement toward not only a transnational film industry but in fact a global cinematic style. His chapter seeks to utilize the background of director/director of photography teams in the early decades of cinema to explore several such creative collaborations at work today. Pairings that will be discussed in depth include Australian-born Christopher Doyle's iconoclastic work with Hong Kong Chinese director Wong Kar-wai, British cinematographer Anthony Dod Mantle's equally groundbreaking photography on the films Danish director Lars von Trier, and the extraordinary long-term collaboration between cinematographer Roger Deakins and the Coen brothers, dating back to *Barton Fink* in 1991. By analyzing the real contributions made by both director and director of photography in each of these pairings, this chapter will examine the collaborative nature of the cinematic art, illuminate the relationship between narrative, drama, and visualization, and above all, explore the ways in which the transnational motion picture industry has generated a distinctive visual style, emerging from a dialectic synthesis of national cinemas.

Patrick Keating's focus is on the phrase, "painting with light," and leads our attention to media interactions. Cinematographers have been comparing their work to painting for decades. Some cinematographers take the analogy with painting very seriously indeed, argues Keating, proposing that the careful study of painting is essential to a deeper understanding of light and composition. His chapter considers the role that the analogy has played in the theories of three prominent cinematographers: Néstor Almendros, Henri Alekan, and Vittorio Storaro.

Peter Rist discusses the relationship between cinematographer Mark Lee Ping-bing and director Hou Hsiao-hsien in the historical context of Taiwanese "new wave." Rist's particular focus is on how the cinematographic techniques of long shot and long take were formulated by the collaboration of the two. According to Rist, there is an interesting documentary on Mark Lee Ping-bing—*Let the Wind Carry Me* (2009), which concludes with Lee and Hou making *The Flight of the Red Balloon* together in 2007. The title of the documentary is indicative of both the direction in which Lee was moving at the time—a free-flowing style—and the "flight of the red balloon." Lee states that after working with Hou for fourteen years from *A Time to Live, A Time to Die* (1985) until *Flowers of Shanghai* (1998) on five of the director's eight feature films they became free to experiment stylistically. One of the aspects of this freer expression is the use of camera movement. Hou was a key figure in the early 1980s Taiwanese "new wave," and two shared stylistic traits of this movement were shooting at a distance from the action, in long-shot scale, and in extended, long (duration) takes. Hou was arguably the prime mover in this, and during this fourteen-year period, Lee participated in Hou's increase in average shot length (ASL) from 24 seconds (*A Time to Live, A Time to Die*) to 158 seconds (*Flowers of Shanghai*). Rist argues that although Lee wasn't necessarily responsible for the extension of Hou's long take style, he was clearly involved in increasing the amount of camera movement, from 23 percent of the shots in *A Time to Live, A Time to Die* to 82 percent of the shots in *Flowers of Shanghai* Interestingly, according to Rist, whereas ASL decreased over Hou/Lee's next four films together (from *Millenium Mambo*, 2001 to *The Flight of the Red Balloon*), the percentage of shots including camera movement stabilized, and even increased to 98 percent on *The Flight of the Red Balloon* Rist examines the effect of Lee's interest in camera movement in his work with other directors, especially the great Tian Zhuangzhuang on his remake of a "classic" Chinese film, *Springtime in a Small Town* (2002), and devotes a large section of the chapter to Lee's participation in Hou's stylistic homage to French Impressionist paintings (and films) in *The Flight of the Red Balloon* and Lee's subsequent collaboration with French director Gilles Bourdos on the award-winning *Renoir* (2012), for which Lee received Best Cinematography nominations from both the American Society of Cinematographers (ASC) and the French film industry, Césars.

The last three chapters engage with the conflicting issues around digital technology and its effects on cinematography. Carlos Rojas turns to the films of Fruit Chan and examines the cinematographic materiality of film and digital video. For Chan, Rojas argues, film connotes a sort of material specificity wherein the conditions of production remain tangible within the finished product, while DV is associated with a process of abstraction that permits the user to bridge vast geographical distances. Whereas the indexical quality of film offers a direct link between the cinematic representations and the original visual scenes that produced it, the link between reality and representation in digital media is much more mediated, as visual stimuli are first translated into data before being translated back into visual images. Examining *Made in Hong Kong* (1997), Chan's first independent film, and *Public Toilet* (2002), his first digital video film, Rojas discusses how each work deploys distinctive qualities of its own cinematic medium. At the same time, Rojas conducts an allegorical reading of these films in terms of the national and transnational position of Hong Kong itself.

Examining the complexity of the transnational collaboration between Spanish director Luis Buñuel and his preferred director of photography, Mexican Gabriel Figueroa, Ceri Higgins argues that their films probed and revealed social and psychological contradictions and conflicts in narratives and characters to make an incisive critique of the sociopolitical contexts within which they made their films. Their creative partnership established a visual and technical model that was to influence subsequent filmmakers in Mexico to the present day. She extends her argument on the transnational, Gothic visual narrative to its subsequent development by filmmakers, focusing primarily on Guillermo del Toro, who with his collaborators, employs contemporary digital technologies to take the Gothic to new levels of expression.

Julie Turnock points out problems of digital convergence by arguing that cinematographers and special visual effects artists are both partners and adversaries. Professional discourse on the relation between special visual effects and cinematography has lately become heated, with voices in both communities grumbling over Cinematography Academy Awards given to work on *The Life of Pi* (2012) and other films with extensive digital environments. DP Christopher Doyle spoke out against the *Pi* Oscar saying, "it's a f—ing insult to cinematography," and more, "It's not cinematography." Effects artists were equally displeased, but because they did not feel properly credited for the extent of their contribution on the film. From an industry standpoint, both sides fear the nearly fully animated environments in this film and others blur the line between long-standing production (and award) categories. Academics might wonder, Turnock claims, are these films instances of the much-promised digital convergence—converting all images to computer code to allow greater ease of flow across imaging and transmission platforms—scholars (Jenkins, Bolter and Grusin, Manovich) have been promising for nearly twenty years? Certainly effects-heavy films that are lit by computer algorithms, shot by virtual cameras on green screen sets, and color-corrected pixel by pixel would seem to be clear examples of the digital convergence these scholars describe. However, Turnock points out that within actual industrial production, the situation is less clear. Turnock argues that what the particular rhetorical dispute between cinematographers and effects artists helps reveal is that in the cinema, and most markedly in special effects-heavy cinema, much of moving image computing remains stubbornly un-converged in actual filmmaking practice. Digital technological convergence, if possible at all, faces two large interrelated hindrances: institutional territorialism and proprietary economics. Analyzing professional practitioner rhetoric on recent special effects-heavy productions in *Gravity* (2013), Turnock places these films within the historical economic context of special effects production and labor. Describing these largely discrete sectors of production and economics have important consequences to destabilizing the smooth and systematized Hollywood business rhetoric of digital convergence as a subset of conglomeration and synergy. The digital material captured or generated by one company's proprietary pipeline is rarely if ever easily interoperable with another. The hyperspecialization each system requires engenders the very mistrust and misunderstanding the professional rhetoric points up. Finally, Turnock argues that convergence is not a technological condition of the digital. Rather, it is the positivized corporate rhetoric of vertical integration. By understanding digital convergence as a rhetoric, we understand there is no such thing as "the digital," but many digitals with incompatible operating systems, pipelines, and software. According to Turnock, this separation is unexpectedly to the benefit of conglomerates, serving to separate teams of workers, rather than foster cooperation or solidarity.

NOTES

1. "The Founding Fathers," *American Cinematographer*, 85, 8 (August 2004). http://www.theasc.com/magazine/aug04/founding/page2.html. Accessed on March 11, 2014.
2. James Wong Howe, "Upsetting Traditions with 'Viva Villa,'" *American Cinematographer*, 15, 2 (June 1934): 64, 71–72.
3. James Wong Howe, "Lighting," *Cinematographic Annual Volume Two 1931*, ed. Hal Hall (Hollywood, CA: The American Society of Cinematographers, Inc., 1931), 47–59.
4. Emery Huse and Gordon A. Chambers, "Eastman Supersensitive Panchromatic Type Two Motion Picture Film," *Cinematographic Annual Volume Two 1931*, ed. Hal Hall (Hollywood, CA: American Society of Cinematographers, Inc., 1931), 103–108.
5. Hal Hall, ed. "Improvements in Motion Picture Film," *Cinematographic Annual Volume Two 1931*, (Hollywood: American Society of Cinematographers, Inc., 1931), 95, [93–102].
6. Michelle Aubert and Jean-Claude Seguin, *La Production Cinématographique Des Frères Lumière* (Paris: Librairie Du Premier Siècle Du Cinéma, Bibliothèque du film, Editions Mémoires de cinema, Diffusion, CDE. 1996), 72.
7. About this exhibition, see Daisuke Miyao, "Lumière!: Le Cinéma Inventé," *Nineteenth-Century Art Worldwide* 14, 2 (Autumn 2015). http://www.19thc-artworldwide.org/autumn15/miyao-reviews-lumiere-le-cinema-invente.

8. Gabriel Veyre, *Opérateur Lumière: Autour du monde avec le Cinématographe Correspondance* (1896–1900) (Lyon: Institut Lumière/Actes Sud, 1996), 275–276. Translation by the author.

9. Ibid., 272. Translation by the author.

Chapter 1

When the Camera Is Cinematographer

The New Documentary Aesthetics of Leviathan

Lindsay Coleman

The 2012 documentary *Leviathan* was nominally directed by Lucien Castaing-Taylor and Véréna Paravel, and is a graphic depiction of the physical rigors endured by fishermen on a trawler as they go about their work. Rather than always operating themselves the two often locked cameras onto their subjects and filmed them going about their work. Equally, they would periodically place cameras in positions such that they would be unable to perceive what they were filming but would instead imagine it. Hence, while the film is a powerful, visually arresting document, the question of who was its author cannot be simply answered, or perhaps answered at all. It is, to a degree, authorless. Castaing-Taylor comments:

> One thing that we did want to do after our previous films was to work in a more reciprocal way with our subjects, to breakdown the subject-filmmaker distinction. So after the first trip, we wanted to find a way to get closer to the ocean, the fish and the fishermen's experience. We started using small DSLRs that could attach to the fishermen's bodies like on their heads, their wrists or their chests.[1]

To a fascinating degree the camera technology used to both digitally capture the work of the fishermen and record the sounds of their toil and that of the environment around them may well be credited as the most influential aesthetic determinant of the film, more so than the directors, who disavow the technical knowledge and skill one might expect in a cinematographer/director, or the bodies to which these cameras were attached, the anonymous fishermen who received no specific direction. Academic Ohad Landesman observes that the documentary in fact replicates a fly's 360-degree vision, comprised of partial angles and view. The "dehumanized camera is impartial, but it is also closer, not only to the textures and 'feel' of men, birds, fish, water, and machines, but also to their sounds." This leads Landesman to conclude that the perspective is one seemingly disembodied, so seemingly divorced from the work of a given operator, a clear inquisitorial perspective.[2] This chapter will argue that it is, intriguingly, the technology itself, through both its strengths and limitations, which constructs the film's unique aesthetic, as well as providing an equally unusual authorial perspective.

Lucien Castaing-Taylor and Véréna Paravel represent not just documentarians and cameramen in their work, but also are specifically documentarian theorists. I was lucky enough in 2003 to spend a number of days speaking with Lucien, and during this period he was erudite and passionate on what he believed ethnographic and documentary film could express, a far cry from the more dry style he frequently observed in documentaries of the period. As far back as the 1990s he has articulated the following:

> But what if film doesn't speak at all? What if film not only constitutes *discourse about* the world but also (re)presents *experience of* it? What if film does not *say* but *show*? What if a film does not just *describe* but *depict*? What, then, if it offers not only "thin descriptions" but also "thick depictions"?[3]

What this statement represents is a rejection of the kinds of discursive documentary wherein, through carefully presented testimony, and carefully contextualized footage, a description of phenomena, complemented by a description of how to consume it, is presented to an audience. One might perhaps think of the polemical works of Michael Moore, or perhaps the talking heads of a film such as *The Gatekeepers (2012)*. Here, editing and camerawork are tightly constrained by the explicit focus of the director, the rhetoric of the film working on the basis of how the audience is lulled into the interpretive patterns deemed optimal by the director. Michael Moore will cut for humor, simultaneously carefully evoking a "fly-on-the-wall" aesthetic, even as he carefully walks the viewer through his interpretation of the events being presented. In *The Gatekeepers* the camera becomes an unblinking eye, the interviewer's questions symbolically pinning the subjects in the frame as myriad thoughts and impulses race across their features in response to each new provocative query. Naturally, I would argue, there is nothing inherently wrong with this style, but Castaing-Taylor argues that it strips the documentary genre of its expressive potential.

In the documentary *Leviathan*, directed by Castaing-Taylor and Paravel, the viewer inspects, in explicit digital camera detail, the grinding and harsh realities of working as a fisherman. They witness fish spilling onto the deck, birds hovering, water splashing out and onto the deck, the straining muscles of the men involved. Yet Castaing-Taylor remarks that "the film exists before interpretation." Put simply this means that, in his estimation, there is no rhetorical technique which might pin down his and Paravel's perspective on this labor. The cuts do not obviously emphasize the humor of a given moment, its political import, or a specific interface between the director and their subjects, as one might find in the talking heads of *The Gatekeepers*. Rather the "purpose was to give people a very potent aesthetic experience, to give them a glimpse into a reality that they haven't had first-hand—a protracted, painful, difficult, visceral, profound embodied experience. . . . Our desire was simply to give an experience of an experience. . . . It's an 87-minute experience of being at sea, both metaphorically and literally."[4] In this context, "experience" might be thought to mean a stream of information and sensation, one which we might need to analyze and interpret at a given remove from the film, equally as our daily sensations arrive to our brain via our nervous system and are gradually distilled and interpreted. While Castaing-Taylor affirms that the film eschews obvious editorializing, he is similarly affirming that we, the audience, are likely as capable as any filmmaker of offering our own editorial application to the sights and sounds of *Leviathan*.

By Castaing-Taylor's account, he and his codirector wish to avoid "an abdication of aesthetic, intellectual and political responsibility, because it is reducing the world to something that the filmmaker is pretending to be able to give you certain pronouncements about, to edify the audience."[5] Certainly this would seem to coincide with their anti-Moore approach, a decision to focus on the work, its obvious hardship, rather than its sociopolitical import. Absent is the Moore-esque narration: "These poor guys work for minimum wage based on laws created by politicians who don't even know they exist." However along with disavowing any distinct rhetorical agenda, these filmmakers equally disavow any technical mastery whereby they captured these visceral scenes of toil. Observe their exchange below:

> *VP:* I have to tell you first that we are a complete technophobe and very spastic.
>
> *LCT:* It's true, she is not being humble.
>
> *VP:* When they asked me about these technical questions about my last project I couldn't help but being repulsed by it. (laughs)[6]

The two, in short, confess to a very limited knowledge of the tools by which they captured their footage. Indeed, much of the film *Leviathan* is arguably marred by artifacts such as digital noise, exaggerated colors, especially orange, and assorted issue with the digital capture system they used, the GoPro HD. Remarkably, this same commercial camera is notorious for being prone to glitches. Examine below an assorted array of such issues, offered by an amateur GoPro user to other prospective users:

> Never take the card out! I don't really consider MICRO SD to be very "pro." Removing the card over and over can only add probability for failure. Instead, turn your camera on, plug in your mini usb and let the camera mount as a hard drive as much as possible.[7]

With this incredible array of issues, and given the filmmakers own avowed lack of technical skill, the question must be asked—in attempting to create an immersive experience why did they opt for the GoPro? Why did they preserve all of the flaws of their chosen camera system in their final edit of the film, especially given that such an attention to aesthetic matter would have likely been the area which they would have been most qualified to intervene with the footage. If they were arguably hamstrung during production, why highlight the flaws of their system even after their work during post?

To answer this perhaps we must look at the camera itself. While there are many versions of the GoPro, here are some recent specs on the camera. The technological profile the camera represents is one where its smallness, toughness, wide-angle lens and broad exposure range appeal across a wide variety of shooting scenarios. Originally a 35mm-film camera it eventually transitioned to a camera capturing silent, 10-second digital videos. This was then followed by a transition to a 3-megapixel camera that contained an SD card slot and sound. Its lens now features an extraordinary scope of 170 degrees.[8] These new features of the camera distort color, sound and the visual plane, and as such would not seem to commend the camera's ability to capture "reality," rather perhaps a heightened and intensified version of the same. Academic Eirik Hanssen points out that "the digital color aesthetic of *Leviathan* challenges our expectations of naturalness in ethnographic representation. The stylization and fluorescent glow of the lighting design is instrumental to the experiential effect of the film. Again, this is not human vision, but technological vision."[9] In choosing the GoPro specifically Castaing-Taylor and Paravel were also opting for a piece of equipment which by its own inventor's definition and the opinion of professional cinematographers is inherently limited and simultaneously graphic in its potential to capture action. The use of "GoPro footage has itself become something of a cliché today; the cameras have become as ubiquitous in certain realms, such as outdoor sports, as cell phones have become in general daily life. . . . We have to acknowledge the degree to which the GoPro, in concert with the distribution apparatus of 'social media' that is central to the camera's astonishing commercial success, has become a constitutive a priori of experience today."[10]

Fascinatingly, this suggests that the technology itself, by dint of marketing, or perhaps its penetration of such a range of markets—sport, feature films, documentary, diarizing on social media—has managed to plausibly make its case as an ideal medium in which to convey "experience," despite its obvious distortions of our visual reality.

By objective measure a mixed bag, even exempting its previously noted tendency to freeze up, the GoPro possesses a litany of quirks, some visual positives, and others visual negatives. In the view of cinematographer Eric Zimmerman, the GoPro possesses impressive automatic exposure, meaning skin tones balance easily. Similarly "in terms of electronic noise and blown highlights, it held up very well." With care, even with its small sensor, impressive and professional results are still possible, given a talented cinematographer and careful use of the camera's features. "Since you only have Auto Exposure, you need to know when to use the Spot Metering setting and when to stay with the Center Weighted mode. In bright light you will get a strobe look with fast action. The lack of a viewfinder isn't really a problem since there is a small external video monitor that is available as an option." On the negative side, Paul Maibaum singles out the GoPro's user control buttons. "When the camera was inadvertently left in the still position instead of video, there was no obvious warning light," "The other drawback is the inability to handle extreme vibration." Peter Pilifian, another cinematographer, points out that the contrast of the image is frequently too extreme, this being the camera's default setting.[11] The camera itself would not seemingly recommend itself to incisive immersion, as found in an ethnographic film such as *Leviathan*, coming as it does from the world of extreme sports and throwaway moments of visual flair. Here is the camera's inventor discussing its inception:

> How did the idea for GoPro originally come about?
> I'd had the idea for a wearable camera kicking around in my head since the late 1990s, but I first really started developing the idea back in 2002 while on surf trip in Australia with a couple of friend. . . . I'd kill for some GoPro footage of that trip! Ironically, that trip is what fired me up to come home and finally start GoPro to create "the invisible camera," a wearable camera so convenient that you forget you've got it on.
> When thinking back about the first time this product idea came to mind, what one memory stands out? Every time one of us would get a sick barrel, we'd say to each other: "Agghh! If only we had a camera!" Every surfer knows that feeling![12]

This enthusiastic statement, to my view, sums up much of what might be called the camera's accidental success outside of its original inception, namely for nonprofessional cameramen to record the exhilarating moments of their downtime. The camera can be used for many things, yet more often than not this fact is down to the skill and counterintuitive choices of those camera-people—feature filmmakers and documentarians alike—who opt to push this camera so far beyond what it was originally intended for as a digital capture device.

Converse to capturing a "sick barrel," the Sensory Ethnographic Laboratory (SEL) at Harvard University, the thinktank from which *Leviathan* originated, focuses on "harnessing perspectives drawn from the arts, the social and natural sciences, and the humanities, the SEL encourages attention to the many dimensions of the world, both animate and inanimate, that may only with difficulty, if it all, be rendered with propositional prose. Most works produced in the SEL take as their subject the bodily praxis and affective fabric of human and animal existence." [13]

In essence the use of the GoPro on *Leviathan* was to explore concepts which would find easier representation with the patient eye of perhaps Jean Rouche, with his 16mm Bell and Howell camera, the ideal device to evoke what is known as *camera stylo*, the fluid and mobile camera whose ability to penetrate the action of a scene was essential in Rouche's early era of ethnographic filmmaking. The GoPro, in contrast, was designed to be locked to a body, and the perspective limited to that of the participant, with little by way of editorializing occuring as the camera did not move, at least not in the sense of being consciously operated. Such a moment is clearly evidenced in *Leviathan* as a fisherman stands, the camera attached to him, and watches equipment move back and forth, back and forth. There is no master shot, not camera zipping from one element of the shot to the next. It is static, its perspective essentially limited. In fact, the openness of *Leviathan* is so open that the film becomes downright mysterious. The withholding of information is not only a holding back of facts; it is also visual. The filmmakers have refused any conventions of spatial orientation, such as master shots, continuity editing, and so on; likewise, there is no identification of people, place, or activity. Their predilection for close-ups, partial views, and unconventional camera angles make it difficult to know what one is looking at much of the time. Upside-down seagulls, nets of fish, open portholes—these are all unfamiliar sights that come to us as sound and color before they are objects. Many of the shots tend toward abstraction, with remarkable visual design, movement, and coloration; and the intensity is furthermore enhanced by a multichannel soundtrack composed of recorded and produced sounds.[14]

This approach would seem radical, yet doubly so given that it would indicate the excision or withholding of information, the choice to obstruct the camera's movement, to limits its power to clarify relationships, to successfully map out the space individuals move through. This would seem to find further confirmation when considering a near simultaneous work of the Sensory Ethnographic Lab, *Manakamana*.

Stephanie Spray and Pacho Velez's *Manakamana* (2013) is a film about pilgrims in Nepal who ride a cable car together to their destination. The pilgrims sit beside one another, in a single shot which clearly maps out the internal space of the cable car and the subject's orientation to one another. Theorist Irina Leimbacher's writes on the film:

> As a viewer we are not sure who or what is looking, how present the recording apparatus is to the subjects, and what their engagement is with the camera

> and the people behind it, if any. It is only through subtle hints in the texture of the image and the comments of the passengers that we slowly come up with hypotheses.[15]

The tools to formulate said hypotheses are all provided to the viewer: objective framing, a finely textured 16mm image, a take rolling in real-time of the subjects' interactions. These interactions may vary, from boredom, to utter detachment, to fondness, to mildly theatrical gesture, depending on how conscious they are of the camera. Stephanie Spray's notes that "later, viewing the footage, I realized that the duration of the trip would allow for intimate exchanges to unfold." Specifically, this involved her "knowing that a 400-foot roll of film at 24 frames per second would be roughly equivalent in cinematic time to the duration of the cable-car trip." Equally, on a spatial level, the frame of the film proscenium echoed the framing which the cable-car windows allowed for the view outside. Geometry, consistent spatial logic, the sense of individual moving through a clearly delineated, consistent-yet-limited space.

In the service of this the filmmakers mapped out the exact position of the tripod so that, regardless of the day, they could be assured of always placing the tripod in the same exact orientation to their subjects. Equally the lens was at a consistent height. They "recorded sound using a shotgun stereo microphone encased in a zeppelin on a boom pole and recorded the sound on a two-channel sound recorder, the Sound Devices 702." This meant that, unlike *Leviathan*, they were keeping the sound recording and the filming of the images separate from one another, using distinct devices. Unlike Paravel and Castaing-Taylor the filmmakers consciously employed a particular stock to shoot in 16mm, which in turn they knew would be scanned specifically for eventual digital projection. The stock they selected had "a great exposure latitude," especially given their keenness to capture in detail their subjects and the colorful background environments passing them in the background. Similarly, given their desire to keep both foregrounds and backgrounds in crisp focus, they chose 16mm stock, which served their ends admirably. They eventually enjoyed a 2K DCP.[16] In brief they wished for a camera with color range specific to and accurate to their subjects, and in turn their environment, an overall sharpness which might help audiences follow the transition of the subject through their environment as they transverse the cable car on one axis, all the while interacting with another. The camera is always aware of where and how it is moving through space, and how objects are moving in front of it.

As Castaing-Taylor admitted, in contrast to *Manakamana*, they often filmed without looking through a viewfinder, without framing what became their final image, particularly off the side of the boat. Equally the fishermen with cameras attached to them were not directed in how or what to film.

Verena: They have one of the most dangerous jobs in the world, so you can't give them direction when they're working. It's actually a very repetitive job, but they have to be careful in order to stay alive. If you put your foot in the wrong place at the wrong time—

Lucien: A cable will take you overboard. We didn't even tell them to do what they normally do.[17]

This resulted in an " aesthetic [featuring a] lack of control and intention."[18] Yet this statement is somewhat undermined by the filmmakers subtle sense of what constitutes authorship. In *Leviathan*, rather than the relaxed social interchanges of friends, partners, and acquaintances, the subject/cameramen are concentrating hard on surviving their shifts. They are, according to the filmmakers, worried about their own survival in a dangerous environment. Unlike *Manakamana*, where the subjects were sensitive to the camera's presence, the fishermen's attention was so diverted by their labor, that they were not thinking about being filmed, or about what they in turn were filming. Where then might the intentionality of authorship enter the frame? Simply, to the filmmakers in the attachment of the camera to their bodies, in the movement of the camera being an expression of their intention to make it through their shift. Thus the final aesthetic of *Leviathan* is undeniably a curious one, yet one its directors insist is deliberate and constructed, if not typically. They note:

> [We had a] kind of technological perspective that there is no filmmaking or directorial agency whatsoever, that you can just place these cameras and *let them do the work.* [emphasis added]. There are really four shots in the whole film that were not held by nor tethered to our bodies. All other shots were either held by us or attached to the fishermen's bodies. The shots of the camera going over and underwater were done, GoPro attached on the sticks held by us.[19]

Even as the waves were crashing around them the filmmakers did not know what would happen next, how they or the fishermen, both operators, might resist the power of the waves. As such "the resulting image is a combination of intent and accident."[20] The accident is the split second response to an unknowable environment, the intents the guiding drive to complete a shift on the deck. Paradoxically, despite their avowed ignorance of the technology, this involved some cursory, tactile involvement in the placement or movement of the camera. Castaing-Taylor notes:

> We were interested in capturing this immediate and corporeal subjectiveness; a portrait of fishing from the inside where the cameras will capture the physicality of what it's like to be a fisherman. When the cameras were strapped to their bodies, they couldn't think about filming. They were just working and the footage literally corresponded to every bodily movement that was made.[21]

By their estimation this alone could be thought to contribute to authorship, an authorship where no deliberate choices were made in relation to framing, choreographed movement, or the manipulation of conditions to optimize their shot.

And yet, at the same time, odd elements of genre, and the new digital cinema seemingly seeped into their work. Contrary to appearances, the film was not in fact highly manipulated at an image and sound level. "A lot of people think that "*Leviathan*" is a bunch of special effects, because of the sound design, the color, the assaulting aesthetic; that's all at odds with documentary realism. But everything was recorded with our cameras and we didn't f-ck with the cameras at all. We're definitely beholden to this experience—what emerges arbitrarily—and to the messiness of the encounters." Indeed, frequently the pixelation, noise, and flaws of the image are retained, the arbitrary nature of the camera itself and its various noted flaws equally an element of this "arbitrary aesthetic." Bizarrely however, this arbitrary sense of authorship, as though the camera were randomly being buffeted and coming to rest, recording the entire time, bolsters what would easily be classified as generic images of horror. *Leviathan* is full of blood from dying fish, hacks of fish heads, fish gasping then dying on the deck of the ship. These images, captured via this arbitrary aesthetic are the stuff of horror. "I was shocked by how long it takes a fish to die. I never thought I could feel any empathy for a fish,"[22] says Paravel. To Castaing-Taylor "we talk about horror, or we talk about film noir, or sci-fi films, or romantic films, but there's a lot of bleeding, a lot of slippage between these genres," he explains. "You can see this film as a sci-fi film in certain regards, as a horror film, as an environmentalist documentary, you can see it as an apolitical documentary, as a nature film on a totally different kind of non-anthropomorphic register."[23] The film is not specifically designed, nor is its cinematography, but it this its images, accidental but thematically contrived, may be imbued by whatever genre lens the viewer requires for said themes: toil, death, danger.

What the chapter has so far attempted to capture is the fragmentary nature of the film *Leviathan* and the many contradictions whereby which it was digitally captured. The majority of cameras used, GoPro HD units, were suited to the extreme environs but not obviously suited to the ethnographic format. Little traditional operating occurred on the film, and yet the attachment of the cameras, partial or otherwise, to human beings, was responsible for an aesthetic the filmmakers believe to representative of human experience in such an environment. Simultaneously, to the viewer, the unique position of many of the cameras in an fairly alien environment resulted in images of eco-horror, of dead and dying fish. The need to make sense of these images, arguably, results in the application of a generic lens on the film's visuals, and constitutes a kind of narrative. The fish are hunted, the fish die. All of this with the filmmakers themselves claiming the camera itself "did the work." Equally, they also admitted that the images generated did, in fact, eventually echo their own subjective experience of the filming, their own subjectivity effectively:

> *Lucien:* But also more real and more evocative of what it was actually like to be there. It became less of a spectacle.
>
> *Verena:* Those cameras captured the light in a very special way because of the brightness of the spotlights on the boats juxtaposed by the pitch darkness of being out at sea. It makes you lose your sense of time and place. The aesthetic became more aligned with our physical and emotional experience of being there.[24]

Paravel's comment here is particularly interesting, as it represents the point at which, unconsciously there is an acquiesence to a point of view, to the notion that, whatever a camera may capture, however accidentally it may make a record, there is still some connection back to the individual who chose to place that technology in that situation. The camera enables the claiming of authorship from a remove, authorship as an echo of experience.

The intriguing legacy of *Leviathan* is that it is a work of major influence existing outside of obvious authorship. Ignatiy Vishnavetsky of the A.V. Club notes its influence on major narrative works such as *In the Heart of the Ocean* (Howard, 2015).[25] Yet they reference it for its style which is so distinctive, yet so hard to easily describe, less for its narrative or thematic impact. The film is not conventional ethnographic documentary, nor is it cinema verite. It is, as Castaing-Taylor and Paravel note, an "experience of an experience." Yet, with the contradictory pulls made on any conventional notion of authorship—is it the work of the filmmakers, is it a literal depiction of the work of the subjects and what they see and hear—a safe description of the film at this time is that it is a surrealist record of experience made possible, conceptually and practically, by the existence of the GoPro. This camera, the camera they used the most frequently on the film, permitted for visual articulation of experience in a manner entirely atypical. With no obvious claims to meaning being made on-screen by the subjects, or offscreen by the filmmakers, it is the camera which speaks through its mixture of flaws and merits.

NOTES

1. Chang, Kee, "Q&A With Lucian Castaing-Taylor & Véréna Paravel." *Anthem*, March 1, 2013. Accessed February 21, 2016. http://anthemmagazine.com/qa-with-lucien-castaing-taylor-verena-paravel/.

2. Landesman, Ohad, "Here, There, and Everywhere: Leviathan and the Digital Future of Observational Ethnography." *Visual Anthropology Review* 31, no. 1 (2015): 13.
3. Taylor, Lucien, "Iconophobia." *Transition* 69 (1996): 86.
4. "The Wordless Sea: An Interview with Directors Lucien Castaing-Taylor and Véréna Paravel." By Jason Ward. Ohcomely (2013) Accessed December 20, 2015 (blog), 2013. http://www.ohcomely.co.uk/blog/799.
5. Ibid.
6. Chang, Dustin, "Interview: Lucien Castaing-Taylor and Véréna Paravel on Leviathan and the Possibilities of Cinema." *Screenanarchy*, February 26, 2013, Accessed January 15, 2016. http://screenanarchy.com/2013/02/lucien-castaing-taylor-verena-paravel-interview.html#ixzz35RK3WQvk.
7. Red Head Windscreen Blog, Go Pro Hero Problems: Tips On Keeping Your Hero 3 Working. April 30, 2013. Accessed December 25, 2015. http://www.redheadwindscreens.com/blog/go-pro-hero-problems-tips-on-keeping-your-hero-3-working/.
8. Moritz, Robert, "Guts, Glory, and Megapixels: The Story of GoPro." *Popular Mechanics*, June 12, 2012. Accessed January 7, 2016. http://www.popularmechanics.com/adventure/sports/a7703/guts-glory-and-megapixels-the-story-of-gopro-8347639/
9. Hanssen, Eirik, "'His Eyes Are Like the Rays of Dawn': Color Vision and Embodiment in Leviathan" *Visual Anthropology Review* 31, no. 1 (2015): 21.
10. Pavsek, Christopher, "Leviathan and the Experience of Sensory Ethnography." *Visual Anthropology Review* 31, no. 1 (2015): 9.
11. Geffner, David, "Heroic Proportions." *ICG Magazine*, June 1, 2012, 54–56. Accessed December 15, 2015. http://0097456.netsolhost.com/ICGJuneGoPro.pdf.
12. GoPro with Founder/Inventor Nick Woodman http://design-engine.com/gopro-with-founderinventor-nick-woodman/.
13. Sensory Ethnography Lab, https://sel.fas.harvard.edu/.
14. Russell, Catherine, "Leviathan and the Discourse of Sensory Ethnography: Spleen et idéal." *Visual Anthropology Review* 31, no. 1 (2015): 29.
15. Leimbacher, Irina, "The Sensory Ethnography Lab: Toward a Post-Humanist Cinema." *Film Comment* 50, no. 2 (2014): 38.
16. MacDonald, Scott, "Interview: Stephanie Spray & Pacho Velez," *Film Comment*, April 11, 2014. Accessed December 12, 2015. http://www.filmcomment.com/blog/interview-stephanie-spray-pacho-velez-manakamana/.
17. Chang, Kee, "Q&A With Lucien Castaing-Taylor & Véréna Paravel." *Anthem,* March 1, 2013. Accessed February 21, 2016. http://anthemmagazine.com/qa-with-lucien-castaing-taylor-verena-paravel/.
18. Fazio, Giovanni, "In the ethnographic realm of the senses: An interview with Véréna Paravel and Lucien Castaing-Taylor." *The Japan Times*, August 20, 2014. Accessed December 17, 2015. http://www.japantimes.co.jp/culture/2014/08/20/films/ethnographic-realm-senses-interview-verena-paravel-lucien-castaing-taylor/#.V6MsuLh97IU.
19. Chang, Dustin, "Interview: Lucien Castaing-Taylor and Véréna Paravel on Leviathan and the Possibilities of Cinema." *Screenanarchy*, February 26, 2013.

Accessed January 15, 2016. http://screenanarchy.com/2013/02/lucien-castaing-taylor-verena-paravel-interview.html#ixzz35RK3WQvk.

20. MacInnis, Allan, "Trippy Leviathan is no eco-horror documentary." *The Georgia Straight*, April 25, 2013. Accessed December 2, 2015. http://www.straight.com/movies/375171/trippy-Leviathan-no-eco-horror-documentary.

21. Chang, Kee, "Q&A With Lucian Castaing-Taylor & Véréna Paravel." *Anthem,* March 1, 2013. Accessed February 21, 2016. http://anthemmagazine.com/qa-with-lucien-castaing-taylor-verena-paravel/.

22. Fazio, Giovanni, "In the ethnographic realm of the senses: An interview with Véréna Paravel and Lucien Castaing-Taylor." *The Japan Times*, August 20, 2014. Accessed December 17, 2015. http://www.japantimes.co.jp/culture/2014/08/20/films/ethnographic-realm-senses-interview-verena-paravel-lucien-castaing-taylor/#.V6MsuLh97IU.

23. Anonymous, "Horror at Sea: Lucien Castaing-Taylor on Leviathan." *The Skinny*, October 1, 2013. Accessed December 5, 2015. http://www.theskinny.co.uk/.festivals/edinburgh-festivals/film/horror-at-sea-lucien-castaing-taylor-on-Leviathan.

24. Chang, Kee, "Q&A With Lucian Castaing-Taylor & Véréna Paravel." *Anthem*, March 1, 2013, Accessed February 21, 2016. http://anthemmagazine.com/qa-with-lucien-castaing-taylor-verena-paravel/.

25. Vishnevetsky, Ignatiy, "In The Heart Of The Sea is ravishing and very corny." *The AV Club*, December 10, 2015. Accessed December 17, 2015. http://www.avclub.com/review/heart-sea-ravishing-and-very-corny-229524.

BIBLIOGRAPHY

Anonymous, "Horror at Sea: Lucien Castaing-Taylor on Leviathan." *The Skinny*, October 1, 2013. Accessed December 5, 2015. http://www.theskinny.co.uk/festivals/edinburgh-festivals/film/horror-at-sea-lucien-castaing-taylor-on-Leviathan.

Anonymous, Red Head Windscreen Blog, Go Pro Hero Problems | Tips On Keeping Your Hero 3 Working. April 30, 2013. Accessed December 25, 2015. http://www.redheadwindscreens.com/blog/go-pro-hero-problems-tips-on-keeping-your-hero-3-working/.

Anonymous, Sensory Ethnography Lab Acessed 4 December, 2015. https://sel.fas.harvard.edu/.

Chang, Dustin, "Interview: Lucien Castaing-Taylor and Véréna Paravel on Leviathan and the Possibilities of Cinema." *Screenanarchy*, 26 February 2013, Accessed January 15, 2016. http://screenanarchy.com/2013/02/lucien-castaing-taylor-verena-paravel-interview.html#ixzz35RK3WQvk

Chang, Kee, "Q&A With Lucian Castaing-Taylor & Véréna Paravel." *Anthem,* 1 March 2013, Accessed February 21, 2016. http://anthemmagazine.com/qa-with-lucien-castaing-taylor-verena-paravel/

Design-Engine, GoPro with Founder/Inventor Nick Woodman April 25, 2012 Accessed December 25, 2015. http://design-engine.com/gopro-with-founder inventor-nick-woodman/

Fazio, Giovanni, "In the ethnographic realm of the senses: An interview with Véréna Paravel and Lucien Castaing-Taylor." The Japan Times, August 20, 2014, Accessed December 17, 2015. http://www.japantimes.co.jp/culture/2014/08/20/films/ethnographic-realm-senses-interview-verena-paravel-lucien-castaing-taylor/#.V6MsuLh97IU

Geffner, David, "Heroic Proportions." *ICG Magazine,* 1 June, 2012, 54–56. Accessed December 15, 2015. http://0097456.netsolhost.com/ICGJuneGoPro.pdf

Hanssen, Eirik, "'His Eyes Are Like the Rays of Dawn": Color Vision and Embodiment in Leviathan' *Visual Anthropology Review* 31, no. 1 (2015): 21.

Landesman, Ohad. Here, There, and Everywhere: Leviathan and the Digital Future of Observational Ethnography. *Visual Anthropology Review* 31, no. 1 (2015): 13.

Leimbacher, Irina 2014 The Sensory Ethnography Lab: Toward a Post-Humanist Cinema. Film Comment 50(2):38.

MacDonald, Scott, "Interview: Stephanie Spray & Pacho Velez." Film Comment, April 11, 2014. Accessed December 12, 2015. http://www.filmcomment.com/blog/interview-stephanie-spray-pacho-velez-manakamana/

MacInnis, Allan, "Trippy Leviathan is no eco-horror documentary." The Georgia Straight, April 25, 2013. Accessed December 2, 2015. http://www.straight.com/movies/375171/trippy-Leviathan-no-eco-horror-documentary

Moritz, Robert, "Guts, Glory, and Megapixels: The Story of GoPro." *Popular Mechanics*, June 12, 2012, Accessed January 7, 2016. http://www.popularmechanics.com/adventure/sports/a7703/guts-glory-and-megapixels-the-story-of-gopro-8347639/

Pavsek, Christopher. "Leviathan and the Experience of Sensory Ethnography." *Visual Anthropology Review* 31, no. 1 (2015): 9.

Taylor, Lucien. Iconophobia. *Transition* 69 (1996): 86.

Vishnevetsky, Ignatiy, "In The Heart Of The Sea is ravishing and very corny." The AV Club, December 10, 2015, Accessed December 17, 2015. http://www.avclub.com/review/heart-sea-ravishing-and-very-corny-229524

Ward, Jason. "The Wordless Sea: An Interview with DirectorsLucien Castaing-Taylor and Véréna Paravel." (2013) (blog) Accessed November 4, 2015. http://www.ohcomely.co.uk/blog/799

Chapter 2

Five Functions of Camera Movement in Narrative Cinema

Jakob Isak Nielsen

Camera movements permeate our visual culture to an unprecedented degree.[1] They have a long history within film and television genres but thanks to portable media and digital technology they also populate computer games, 3-D visualizations as well as online videos on social networking and video streaming services such as YouTube and Periscope. This article addresses the functions of camera movement in narrative cinema. Some of the proposed functions may apply to other forms of audiovisual communication but this line of enquiry is for other scholars to pursue.

The most important reason to study camera movement is that it has a profound influence on the way films *look* and the way films are experienced and engaged with by spectators. When we see the effects of camera movement we see an entire image in the process of changing and consequently the narrative and aesthetic functions of the film are likely to change accordingly. Practically since the birth of cinema filmmakers have found occasion to move their cameras—even in the face of staggering technical obstacles. Executing mobile shots entails extra work: laying tracks, following focus, tying the footage into an editing pattern, accommodating lighting schemes so that every second of the shot is lighted in the desired way—not to mention the safety regulations required when using camera cranes or drone cameras. But why do filmmakers bother? One way of answering that question is to look at what narrative and aesthetic functions camera movements actually serve. This is the research question that this article poses.

The article will describe five functions of camera movement that will all be exemplified, including a number of sub-functions. They are informed by extensive viewing performed in order to chronicle the *history* of camera movement[2] supplemented by a select sample: a corpus of between 20 and 30 camera movements were selected within seven categories of movement:

push-ins, pull-backs, arcing shots, handheld shots, booming and aerial shots, follow shots, and autonomous movements. These camera movements were analyzed individually, their functions extracted and assembled into a list of functions for each type of movement. The functions of each of the seven types were compared to each other. Two important conclusions resulted from this work: 1) Although the functions of the various camera movement types were different, there were so many similarities among the various types that it made better sense to organize them as *one* taxonomy of functions rather than for instance a set of functions for each type. 2) Some camera movements have quite specific functions but many camera movements tend to multitask, that is, they perform more than one function either sequentially (shifting from one function to another) or simultaneously (performing several functions at once).

FIVE FUNCTIONS OF CAMERA MOVEMENT

The five functions of camera movement in narrative cinema that will be highlighted here are the following:

1. Orientation: orienting the viewer spatially.
2. Pacing: contributing to the cinematic rhythm of the film.
3. Inflection: inflecting shots in a suggestive, commentative or valuative manner.
4. Reflexive: inviting spectators to engage with the artifice of camera movement.
5. Abstract: visualizing abstract ideas and concepts.

These are not necessarily the *only* functions of camera movements. Elsewhere I have suggested a taxonomy of six functions of camera movement (Nielsen 2007). Excluded here is the function of associating the movement of the camera with the viewpoints of characters or entities in the story world. Others may of course propose alternate or additional functions. However, I believe that the five functions proposed here are all significant and primary. Below is a presentation of each of these five functions including declensions and sub-functions, exemplified with concrete examples.

Orientation

Camera movement can orient the viewer spatially in different ways and it is possible to distinguish between three main parameters: lending more *depth* or *volume* to the image, *directing* the viewer's attention to salient story information, and articulating the *scope* of action.

Using camera movement to articulate depth and volume may be regarded as central to the entire medium of film but it has also been a specific strategy of many filmmakers:

> Of course, even in dollying, as a rule we find it's a good idea to *pass* things in order to get the effect of movement. We always noticed that if we dollied past a tree, it became solid and round, instead of flat. So we used to take buildings with pillars and get a wonderful effect dollying past.[3]
>
> We decided to create a third dimension on a two-dimensional film not only with lighting and composition but also with constantly moving the camera. Since the objects change positions differently in different planes, the camera move reveals the right perspective of the objects in space, thus creating the missing third dimension.[4]

Dwan and Zsigmond emphasize how a two-dimensional plane can come to articulate depth and volume by means of camera movement. Since movement parallax can be read from the *entire* image as a result of "the camera movement-effect,"[5] camera movement will usually supply the image with more depth cues enabling the viewer to make a better assessment of the relative distance of objects in the frame. Layering the image into multiple planes will of course provide better conditions for camera movement to perform this function. Furthermore, camera movement can also allow the viewer to read volume from the dimensions of the objects. This is most often a relatively inconspicuous function, but occasionally it forms the basis of a revelatory function as in this example from *Svengali* where it is precisely *because* of the camera movement that we are given "the missing dimensional information" that was concealed from us before (Image 1).

Garrett Brown (inventor of the Steadicam, the SkyCam as well as many other novel camera supports) and George Miller (director of *Mad Max*, *Lorenzo's Oil,* etc.) similarly address this function but do not only consider it in relation to surface structures of the image:

> When the camera begins to move, we are suddenly given the missing information as to shape and layout and size. The two-dimensional image acquires the

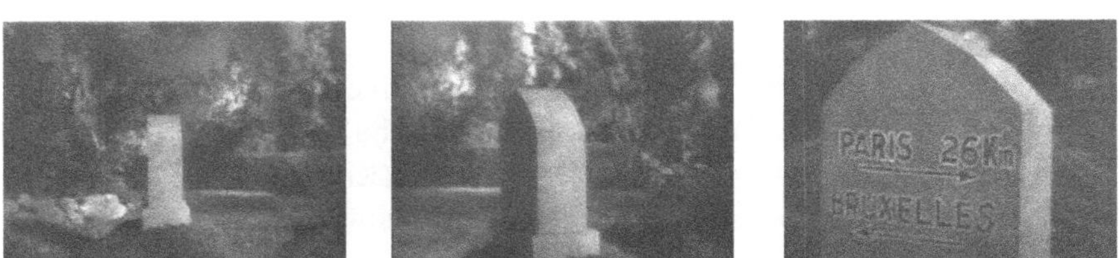

Image 1 ***Svengali* (1931).** What looks like a tombstone turns out to be a road sign.

> illusion of three-dimensionality and we are carried across the divide of the screen, deeper and deeper into a world that is not contiguous to our own.[6]
>
> It's a compulsion of mine to move the camera, and I now know why. It enhances three-dimensionality. It puts you in the space, and if you move the camera the audience becomes aware of the space.[7]

Zsigmond and Dwan's explanations do not *have* to rely on a conception of the camera as a substitute for the movement of the viewer whereas Brown and Miller specifically address this type of spatial orientation in relation to an expressed goal of transporting the viewer into the diegetic space of the movie.

Camera movement can also minimize cues of depth and volume—for instance by swish-panning or otherwise making a blur or "smear" out of the image.[8] At least as pervasive as the depth/volume application is the function of using camera movement to direct the attention of the viewer at salient story information. In this regard the camera functions as a pointer. What Michael Chapman, ASC, has said about cinematography in general is particularly true of this function: "The job of the cinematographer is to tell the audience where to look."[9]

Many different structural types of movement can accomplish this task, for instance lateral tracking shots that keep the viewer's attention on one particular character by accompanying her through a crowd (*accompaniment*). A great deal of editing-in-the-camera that Noël Burch and Lutz Bacher refer to as narrative-expressive camera movement modulates the attention of the viewer through the course of a scene.[10] A case in point is the four-minute long take in William Wyler's *The Letter* (1940) in which the camera repositions to accentuate each development in the duel of words between Howard (James Stephenson) and Leslie (Bette Davis) (Image 2).

Other films direct the viewer's attention with more elaborate movements whose compositional stops reconstitute a whole range of shot scales and compositional units such as two-shots, over-the-shoulder shots and singles (Image 3) as in this conversation between Margot (Gwyneth Paltrow) and Eli (Owen Wilson) in *The Royal Tenenbaums* (2001).

Directing attention can also take the form of *interplaying* two motives, for instance two characters and a symbolic motif as Siegfried Kracauer noted in *Scherben* (1921): The camera moves from a shot of a scarecrow fluttering in the wind to a shot of a window where we see the silhouette of the inspector and the daughter about to initiate a fatal romance.[11]

Of course, direction can extend beyond the frame by *connecting* a motif with another one outside the field of view as in the final helicopter shot of *Picnic* (1955). For instance, that connects Madge (Kim Novak) departing by bus to meet up with her spouse-to-be Hal (William Holden) in Tulsa. Hal is travelling—or rather fleeing—to the same destination by a train that the helicopter shot ultimately brings into view (Image 4). In this case the camera

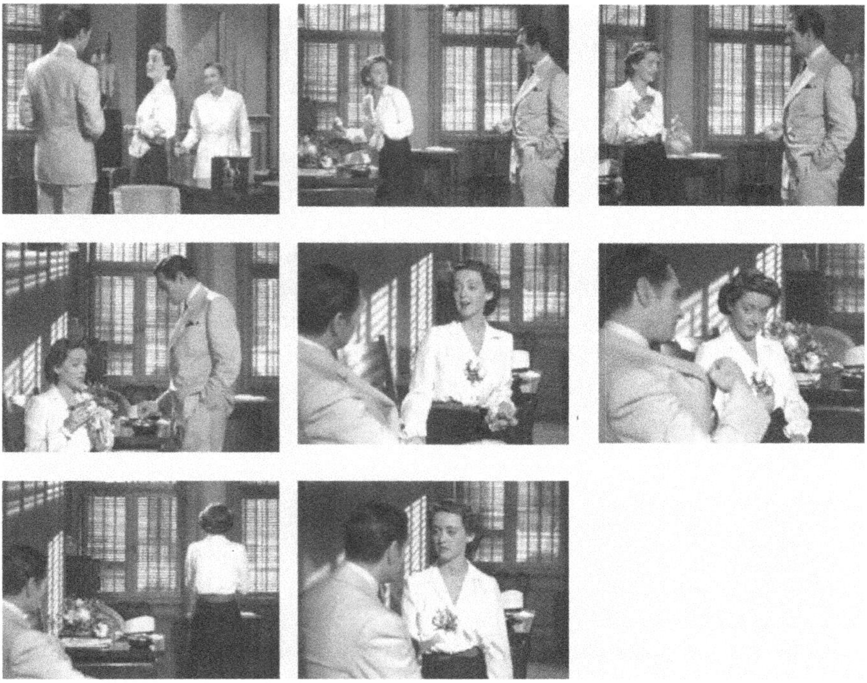

Image 2 ***The Letter* (1940).** Reframing movements accentuate the dramatic development of Howard and Leslie's conversation. *Note:* how this scaled down editing-in-the-camera allows many 'intermediate' positions that would disobey the rules of continuity editing: e.g., the 30-degree rule or the rule of varying field size across a cut.

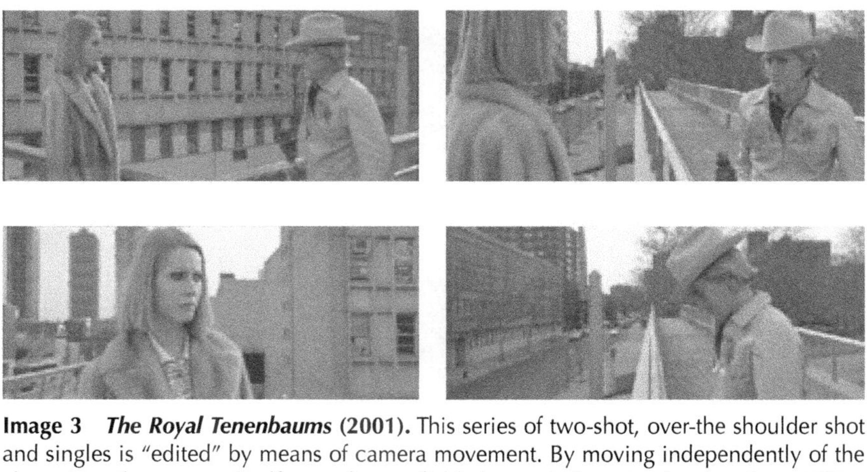

Image 3 ***The Royal Tenenbaums* (2001).** This series of two-shot, over-the shoulder shot and singles is "edited" by means of camera movement. By moving independently of the characters, the camera itself reconfigures field size and character framing thus shifting character-dominance in the course of the shot.

connects the two physically attractive but socially displeased characters thus suggesting a happy reunion in Tulsa rather than actually showing it.

A more subtle way of directing attention is to allow for *planar interplay* during a camera movement such as when the camera follows the agitated Baron Eggersdorff (Gustaf Gründgens) in *Liebelei* (1933) down a flight of stairs passing a statue of a roaring lion on his way or when the two rivals André (Charles Boyer) and Donati (Vittorio de Sica) pass a couple of fencers in *Madame de.* (1953).

The total effect resulting from connecting or juxtaposing two or more motives by means of camera movement can be close to limitless since the possible configurations depend on the number and *nature* of the motives being juxtaposed (Image 5). However, the claim made here is not that the terms *interplay* and *connective* encapsulate the full *meaning* of the *shots* in which they are engaged but simply that they capture the *functional contribution* of *camera movement.*

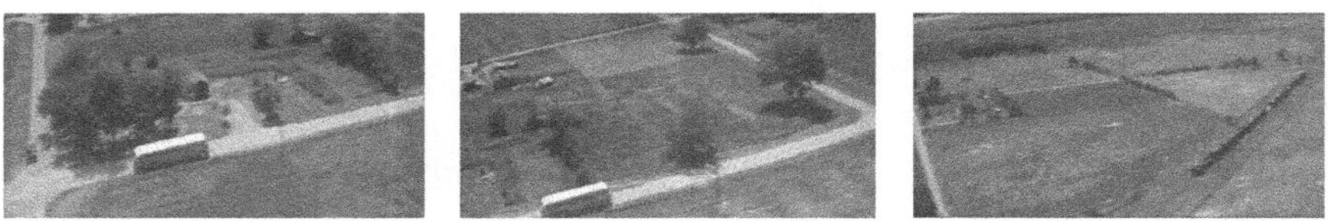

Image 4 The helicopter shot in *Picnic* (1955). Shot by second unit camera operator Haskell Wexler for James Wong Howe. See *Visions of Light* (1994) for Wexler's recollection of the shot and Howe's reaction to it.

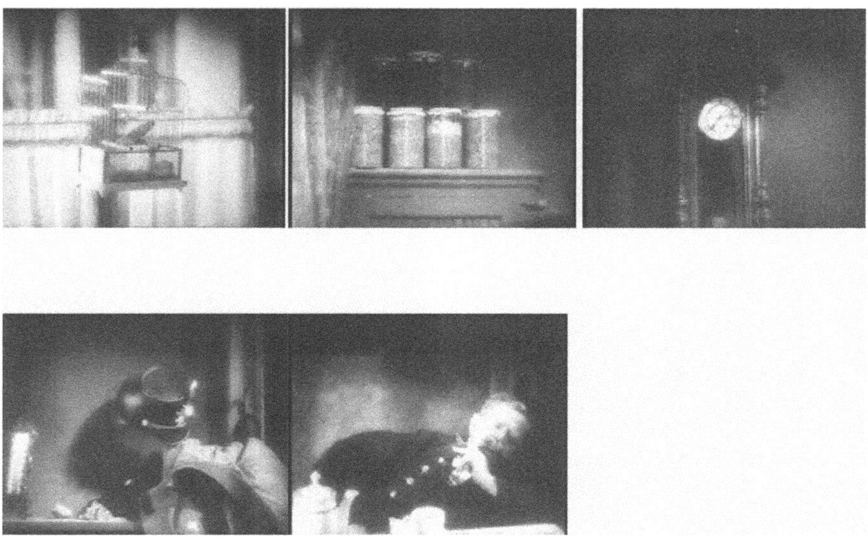

Image 5 The opening shot of *Asphalt* (Joe May, 1929) uses camera movement to line up a number of objects that in unison serve to describe and introduce the household.

Directing the viewer's attention by means of camera movement can be achieved in a multitude of ways. Camera movement can also single out or locate a motif among a variety of possible focal points *(excerpt)*. A minor modality would be the push-in that singles out an action such as Brandon (John Dall) clutching a gun in his pocket in *Rope* (1948), whereas a major modality would be the famous crane shot from Hitchcock's *Young and Innocent* (1937) which locates the twitching eyes of a drummer man in the midst of a crowded room (Image 6).

So pervasive is the function of directing attention that one may ask which camera movements do *not* direct attention? Even camera movements that deflect our attention from a specific action are redirecting our attention (Image 7). Pull-backs that do not emphasize new information at the edges of the frame but simply magnify the spatial layout could be an example. Many of the depth- and volume-enhancing diagonal tracking shots in *Cabiria* (1914) could be another. Many of these shots in *Cabiria* are clearly coordinated according to character movement and particular objects but they are not

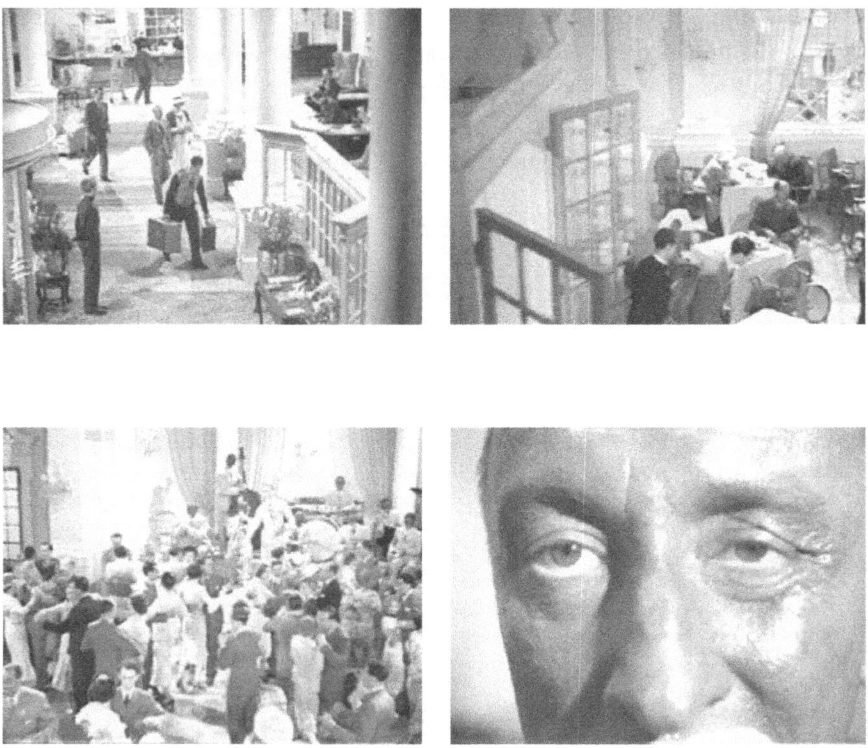

Image 6 ***Young and Innocent*** **(1937).** The famous crane shot from Hitchcock's film, which—within a giant room—singles out the 'twitching eye' that the main characters are looking for.

synchronized with character movement; that is, in many cases the movement is not directed at particular characters or motifs. Instead it is rather the characters that step into the mobile frame at carefully choreographed moments (Image 8). Hence the movement of the camera is not as much directing attention as it is providing a continuously changing playing space.

The last parameter of spatial orientation is *scope*. Here the function is not to direct attention toward specific characters or objects but to establish or re-establish the scope of action. The opening shot of *His Girl Friday* (1940) precisely and economically establishes the Morning Post workspace (fig. 30–32). Many other films—from *Sunnyside Up* (1929) to *Gravity* (2013) use an opening shot to articulate the scope of action but stage these opening shots as virtuoso attractions in their own right (cf. the reflexive function).[12] Other filmmakers will use follow shots to articulate scope. Paul Thomas Anderson likes to use extended follow shots to take "a relaxing moment to watch someone walk through their world."[13] In *Boogie Nights* (1997) the articulation of scope and the connective function are combined when a Steadicam shot follows Jack Horner (Burt Reynolds) walking through his house and yard thus stringing together the members of his peculiar "family." One also finds Woody Allen using long receding tracking shots that integrate his characters into the setting (for instance in *Annie Hall* [1977], *Manhattan* [1980] and

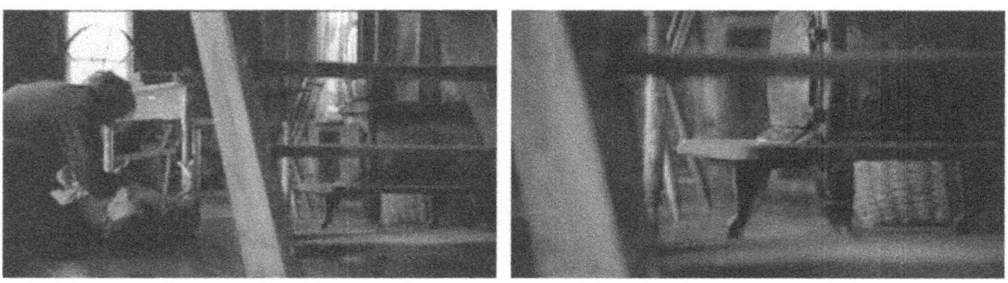

Image 7 ***The Village*** **(2004).** The lateral camera movement and a zoom deflect our attention from the brutal murder. See also Wandering camera—a reflexive camera movement described later in the chapter.

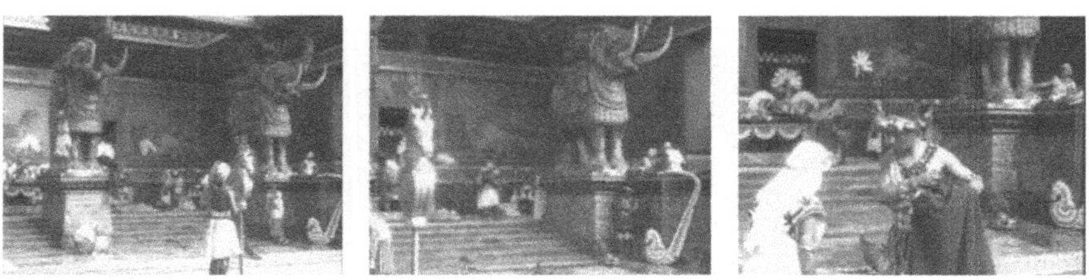

Image 8 ***Cabiria*** **(1914).** The movement of the camera is not as much directing attention as it is providing a continuously changing playing space that the characters can step into or out of.

Hannah and Her Sisters [1986]), in his case often a New York locale thus creating a stronger link between his characters and the spaces they inhabit.

Besides establishing the *spatial layout* of a scene, camera movement can also make a point out of *revealing* scope as when the camera cranes up at the end of *The Good, the Bad and the Ugly* (1966) to reveal the immensity of the graveyard and hence the impossible task of finding the right gravestone (Image 10).

On rare occasions, camera movement *conceals* the spatial scope of a scene as in *Play Time* (1967) where the camera—and Hulot (Jacques Tati)—from their vantage point on the second floor register Giffard's placement amidst the office cubicles below. But as the camera descends along with Hulot on an escalator, the former spatial supremacy vanishes and the scope of the space is concealed from us (Image 11).

The final sub-function to be highlighted is *envelop*. Here the function of the movement is not so much to reveal new information as to envelop a group of characters in a shared space. In *Le Crime de M. Lange (*Jean Renoir, 1936) a pull-back envelops the community gathering around Lange (Rene Lefevre) after the assumed death of their boss Batala (Jules Berry) (Image 12).

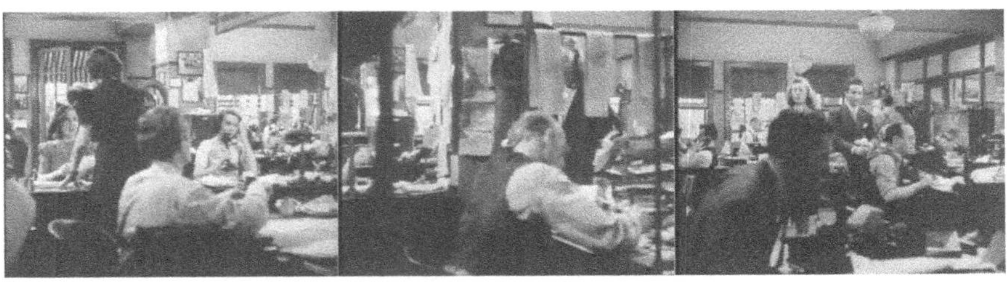

Image 9 ***His Girl Friday*** **(1940).** Establishing locale (read from right to left).

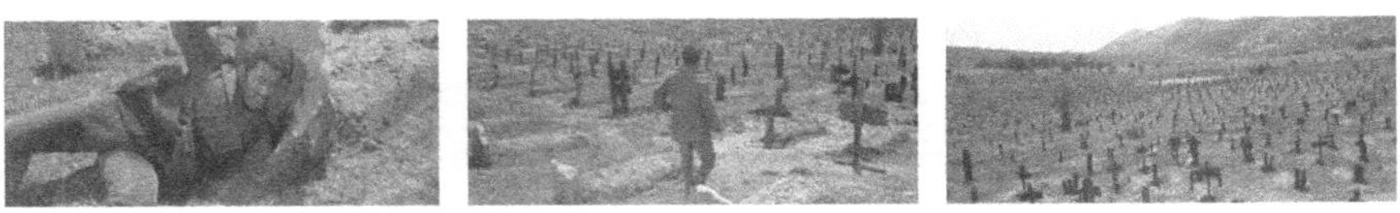

Image 10 ***The Good, the Bad and the Ugly*** **(1966).** A crane movement reveals the scope of the graveyard.

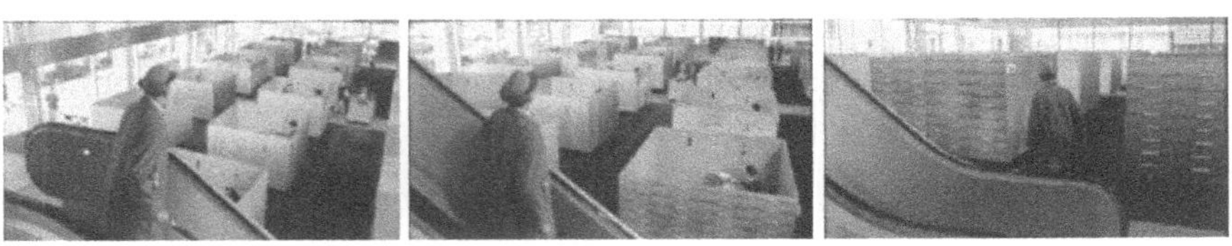

Image 11 ***Play Time*** **(1967).** A camera movement conceals the scope of the space.

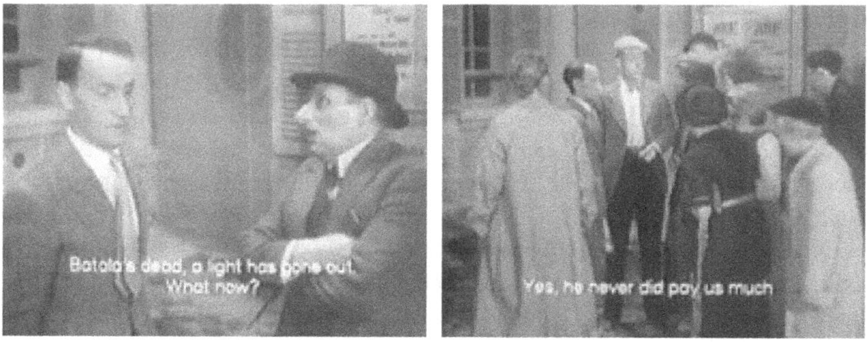

Image 12 **A pull-back envelops the community in *Le Crime de M. Lange* (1936).**

Pacing

Camera movement is a *visual process* as opposed to the visual shift brought about by the straight cut and as such camera movement is not only apt at connecting elements or following action but also very well suited for processual effects.

Camera movements articulate both space and time. Those articulations are integral to the very on-screen effect of "camera movement" and spatial orientation both depends on and is modulated by the time it takes the camera to present the spatial layout. Temporal articulation applies to all camera movements, that is, it is a *resource* that camera movement can make use of to bend a shot to different functional ends

However, it is possible to extrapolate an identifiable function of "temporal articulation" that I refer to as *pacing*. Pacing is a deliberate strategy of "temporal articulation" in which camera movement—for instance a mobile long take or a series of movements — actively contributes to the cinematic rhythm of the film.

Cinematic rhythm is articulated by means of a range of devices: cutting rate, pace of delivery, music, the movement of the players and objects, the fluctuations of shot scale, the duration and "temporal structure" of events shown, their frequency and their presentation. A gun shot takes fractions of a second but may be presented in 10 seconds by means of slow-motion virtual camera movement as in *The Matrix* (1999) while a flight across the Atlantic from New York to London takes about seven hours but may be presented in a four-second montage as in *Snatch* (2000).[14]

The contribution of camera movement to cinematic rhythm is in itself complex. It is of course a question of camera movement velocity but more significantly of the graphic rhythms engendered by the movement of the camera *in relation* to mise-en-scene and the contribution of camera movement to informational delivery. To specify and simplify matters, I will define the function as that of *using camera movement to determine the pace at*

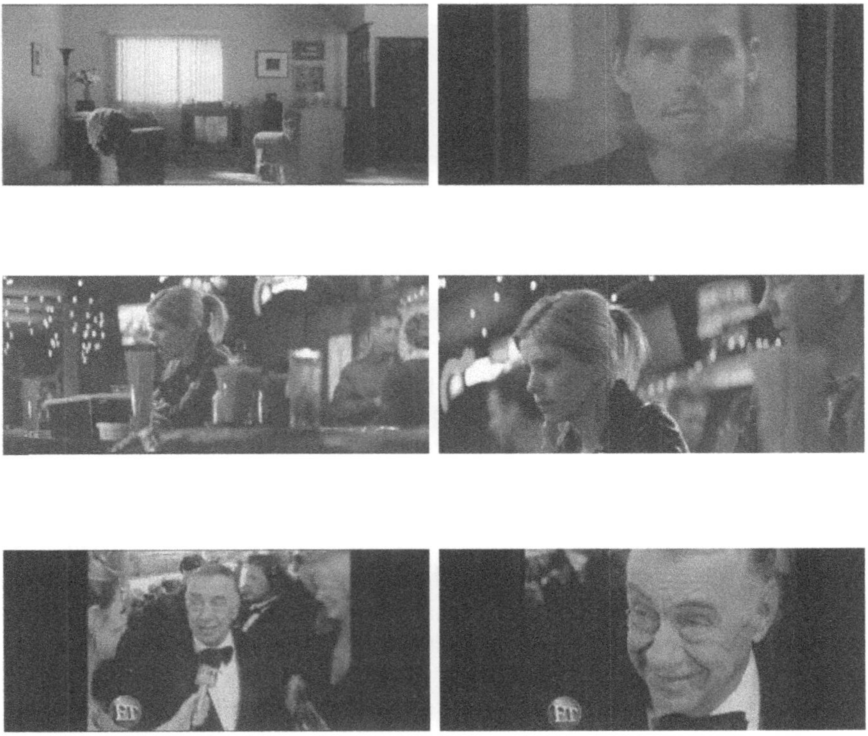

Image 13 ***Magnolia* (1999).** Push-ins are used throughout the prologue and all of Anderson's nine main characters are introduced by means of a forward moving camera or a zoom-in.

which visual information transpires. Information is used here in a broad sense encapsulating everything from graphic details to significant objects. Various films testify that camera movement has indeed been used as a central means of manipulating cinematic rhythm in at least these three ways:

1. Retard. When camera movement is used to deflate or retard the pace at which visual information is disclosed to spectators. The slow and measured camera movements in Carl Th. Dreyer's *Ordet* (1955) exemplify this modality.
2. Cumulate. When camera movement is used to intensify the flow of images to a rising cumulation. The relentless forward tracking shots and zooms in the beginning of *Magnolia* (1999) exemplify this modality (Image 13).
3. Sustain. When camera movement is used to sustain the visual flow of images. The creeping push-ins seen in establishing shots in *Lara Croft: Tomb Raider* (2001) and *Insomnia* (2002) sustain a sense of visual smoothness and fluidity (Image 14).

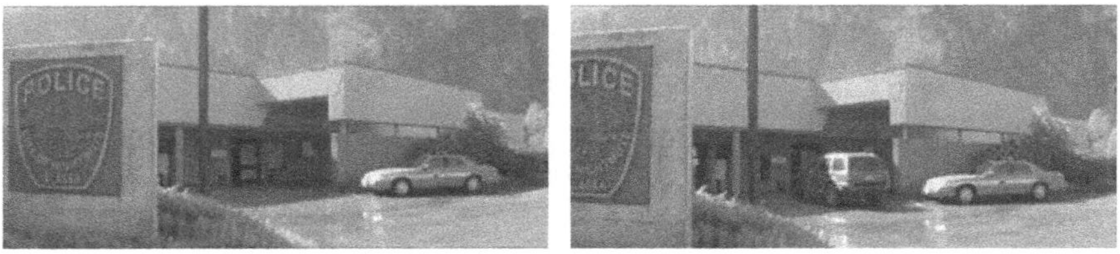

Image 14 ***Insomnia* (2002).** A slow creeping dolly move on a shot with no principal actors sustains a sense of visual flow.

Taken as a whole these modalities emphasize how camera movement can modulate the tempo at which we take in visual information in the course of a film. While these modalities also suggest short-term applications, pacing can also be used by a film as a more generalized strategy of orchestrating visual flow over longer stretches of time: One may specify the meditative and reflective pace of camera movement in Andrei Tarkovsky's *Stalker* (1979), for example, as it traverses the sewers of "the Zone," or the somber, narratively retarding pace of camera movement in Jean Pierre Lefebvre's *The Last Betrothal* (1973), which stretches out the disclosure of objects and characters, the smoothness of contemporary visual syntax or the kinetic and dynamic pace of the continuously reappearing intricate follow shots in Mikhail Kalatozov's *The Cranes are Flying* (1957).

One of the conceptual problems with pacing is that it is often combined with other functions. The above examples indicate that when camera movements are used to determine the pace at which visual information transpires, they often come to communicate particular emotional or psychological qualities. This is sometimes called an expressive function (Bordwell 2005). I use the term inflection here since we are dealing with a slightly wider function.

Inflection

Camera movements can imbue or *inflect* the characters, motifs and spaces that it captures (A→A'). There are at least three key ways in which camera movements can do this. They can imbue shots with narrative significance, they can describe, magnify or bring out "feelingful" qualities, and they can ascribe values to characters, spaces and motifs.

For instance some camera movements not only direct the viewer's attention toward a motif or action but also imbue it with dramatic intensity or narrative significance. The push-in from *Rope* (Image 15) obviously directs the viewer's attention by singling out Brandon's hand clutching a gun within a wide composition but the rapidity with which the camera suddenly breaks into a semi-static scene, the distance it travels and the closeness it achieves also inflects the shot with dramatic intensity.

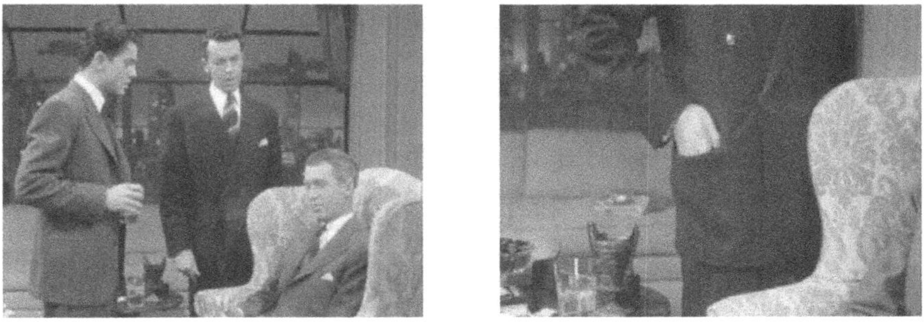

Image 15 Intensified orientation in *Rope* (1948).

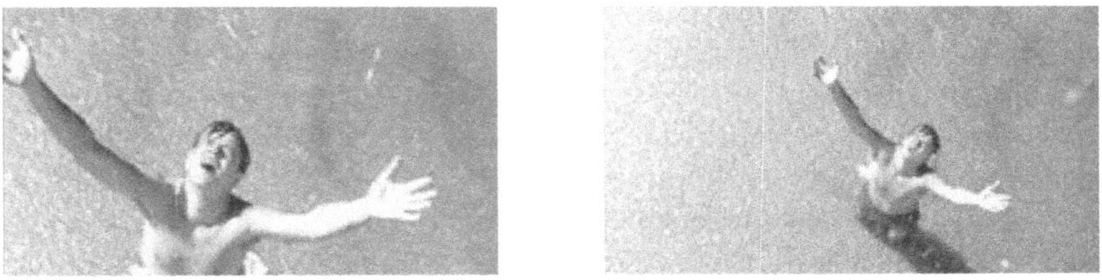

Image 16 *The Shawshank Redemption* (1994). Magnifying emotional and physical outburst.

The ability to describe, magnify or bring out emotional or psychological qualities is sometimes referred to as the expressive function.[15] It is difficult to imagine that the camera can communicate *character* psychology without relying at least partially on the expressive features of the human figure.[16] A classic example is when camera movement *magnifies* a physical or emotional outburst as in *The Shawshank Redemption* where the camera cranes up to visually extend Andy's expressive gesture after he has successfully escaped from prison after crawling through foul-smelling sewers (Image 16).

Camera movement can play a more or less constitutive role in *evoking* or even *bringing out* psychological qualities that would otherwise only have a latent existence in a scene. F. W. Murnau seemed to think that the camera could convey such qualities by itself when writing that the camera should "whirl and peep and move from place to place as swiftly as thought itself, when it is necessary to exaggerate for the audience the idea or emotion that is uppermost in the mind of the character" (Murnau 1928, 90). The push-in has by convention become a strong cue of psychological activity on the part of the character whom we are witnessing: a sudden realization, a flow of powerful feelings or oncoming recollections.[17] Preceding narrative information will often have led us to associate the movement of the camera with a particular psychological activity and beyond this pre-coding the camera move can also be staged in conjunction with a performative and a musical cue.

A forceful way of suggesting psychological activity in the mind of a character takes place in the Danish films *Okay* (Nielsen, 2002). In *Okay* Nete (Paprika Steen) arrives at the hospital to visit her father. We are first presented with a shot of Nete walking into the ward, then a shot that shows us her point of view of the bed that used to hold her father but is now empty; a reverse shot shows us Nete in close-up. Steen's performance, a quiet musical cue and the context in which the shot occurs does to some extent let us in on the thoughts and emotions that she is harboring: Has he died etc.? Nevertheless it is camera movement that makes the most powerful contribution in suggesting psychological and emotional activity (Image 17). Whereas Nete's facial expression does not change in the course of the shot, the camera initiates a Dutch tilt on the close-up of her face and it is primarily the movement of the camera which carries the meaning of the shot by *bringing out* the gradual realization and onrush of emotions and memories that Nete's facial reaction withholds from us.

Camera movements not only inflect and imbue individual characters but also bring out psychological or "'feelingful" qualities in spaces, interactions and situations. When filmmakers broach this topic, they often talk about camera movement's ability to inflect the "mood," "tone," or "atmosphere," of a scene or sequence.

Camera movement will often be used to enhance certain *types of interaction* such as romantic encounters, a scuffle, or an argument by substantiating the type of interaction pictorially: For instance all the way back to the late 1920s—in *Napoleon* (1927) and *Dans le nuit* (1929) for instance —handheld camera movement has been applied to substantiate situations of physical action pictorially whether it be a snowball fight or a skirmish or as in the much later Hollywood film *Body and Soul* (Robert Rossen, 1947) a boxing match.

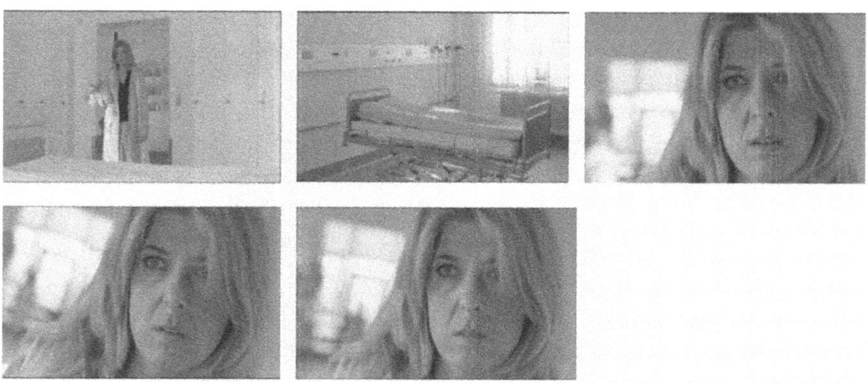

Image 17 ***Okay*** **(2002).** The camera suggests psychological and emotional activity by doing a Dutch tilt on Nete as she sees the vacant bed that is supposed to hold her father. The camera comes up slightly again at the end of the shot.

A camera movement may also serve a function that I refer to as *spatial inflection*. In this case the camera movement is not merely establishing the spatial layout of the scene but is either assigning particular significance or psychological qualities to the space. In *Judgment at Nuremberg* (Stanley Kramer, 1961) there is an arcing shot around the prosecutor (Richard Widmark) as he is presenting his view of the case. As the camera arcs around Widmark's character, the background remains in relatively sharp focus. Stylistically the shot conveys that this is not an ordinary case but a case that calls into question the very setting in which the action takes place—the House of Law. Hence, the arcing shot underscores the prosecutor's comments: "The case is unusual in that the defendants are charged with crimes committed in the name of the law. . . . Therefore, you, your honors, will be sitting in judgment of judges in the dock and this is as it should be for only a judge knows how much more a court is than a *court room*. It is a process and a spirit. It is the House of Law." The arcing shot underscores that line of dialogue in at least two senses: 1) at this stage in the film the camera movement is unusual as is the case and 2) the plasticity that the arcing shot lends to the space underscores the question about the very setting posed by the prosecutor (as well as by the defendant later on): the status of the House of Law (Image 18).

Examples of camera movement ascribing particular values (e.g., positive or negative values) to a character or a space are uncommon but in Milos Forman's *Hair* (1979) one finds arcing shots serving a "celebratory function." During the musical performance of *Aquarius*, the soloist (Ren Woods) is represented in no less than six arcing shots. The arcing shots set her off from the other characters in Central Park as none of the other dancers or performers

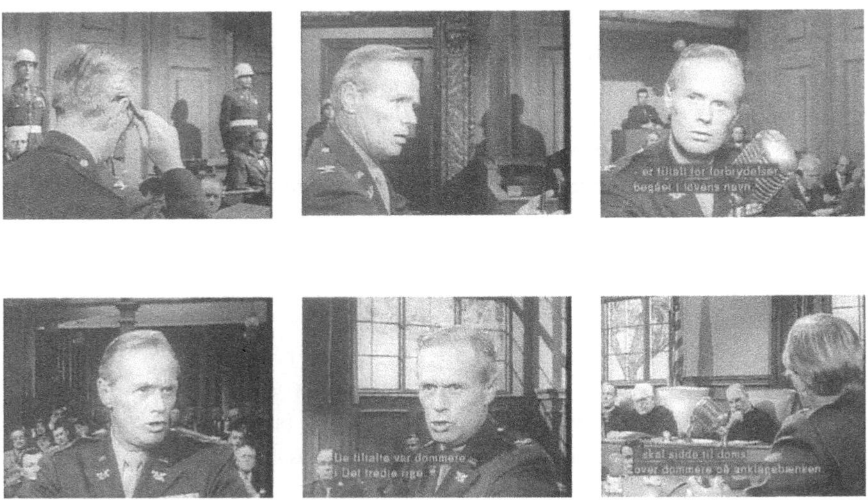

Image 18 ***Judgment at Nuremberg* (1961).** Spatial inflection.

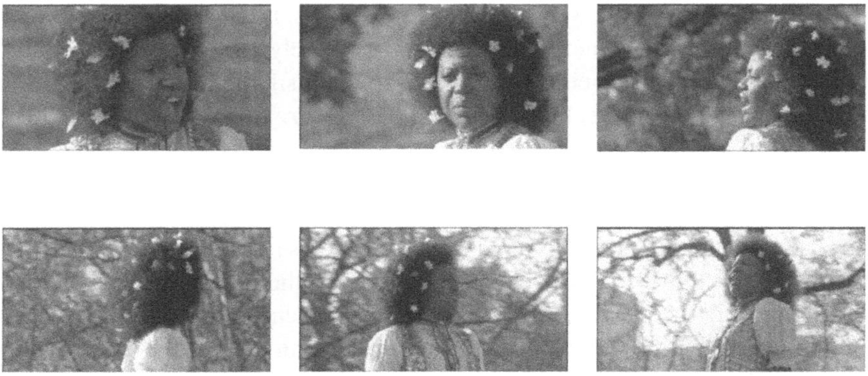

Image 19 ***Hair*** **(1979).** The arcing shots embellish Ren Woods.

are embraced by such a camera move. Other devices help to distinguish the singer—for instance her costume and performance—but the contribution of the arcing shots is significant: First, the camera spirals downward making *her* appear to spiral upwards toward the sky. Hence it sets her off from the other characters on a vertical plane and makes her appear to be spinning in mid-air. Combined with—at times—a low-angle view that makes her sing against the backdrop of tree tops and a blue sky, the arcing shots untie her from gravity and suggest superhuman qualities as if she were a connective link between the firmament and the people of the world—an earth mother (Image 19).

Reflexive

Reflexive camera movements invite spectators to engage with the artifice of camera movement and therefore appeal to specific *types of engagement* divergent from viewers' involvement in the ongoing story.

In meta-film *Hellzapoppin'* (H.C. Potter, 1941) the camera follows Ole Olsen and Chic Johnson as they discuss whether to pick up arrows or sparrows for a particular scene. When they pass a swimsuit model the camera—curiously, it is operated by *projectionist* Louie (Shemp Howard)—is more interested in the female model and keeps returning to her rather than follow Chic and Ole. Chic steps back into frame repeatedly and complains directly to Louie, that is, to the camera. When Chic makes a threatening gesture at Louie, Louie's camera shakes in return! (Image 20)

Although many reflexive camera movements disrupt or subvert the integrity of the fictional world, the term *reflexive* is used here in a broader sense. Reflexive camera movements can be engrossing in many ways; they only bypass the storytelling resources of the medium in order to elicit other types of engagement and appreciation.

For instance camera movements can be staged as *virtuoso attractions* in their own right. Camera movements tend to do this in a number of ways. The move can staged so as to produce or enhance aesthetically pleasing visual rhythms in the structure of the image as is characteristic of many of Max Ophuls' films.[18] One of the ways in which filmmakers achieve this graphically decorative effect is by designing shots so that objects intervene between the camera and the motive at set intervals thus creating pleasing visual rhythms. Another variant is camera movement that allows us to take in large-scale scenery from free-ranging, privileged viewing positions (often helicopter or drone shots). A third is *virtuosity of transport*: how fast the camera can fly, from what heights it can plummet, through what objects it can pass and from what positions it can view the action (often CGI shots). The opening shots of *Lord of the Rings—The Two Towers* has it both ways. The first three aerial shots—separated by dissolves—are mobile "vista movements," but the third shifts to virtuosity of transport as the camera appears to fly straight through a mountainside continuing into the depths of the mountain where it locates Gandalf fighting a Balrog (Image 21).

A less impact-driven "move-through-quality" can be located as far as back as the teens. The best early example is the pull-back in Benjamin Christensen's

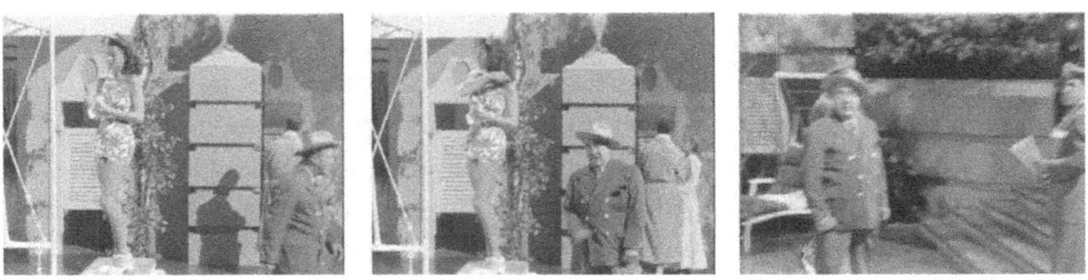

Image 20 The camera draws attention to the artifice of its movement in *Hellzapoppin'* (1941).

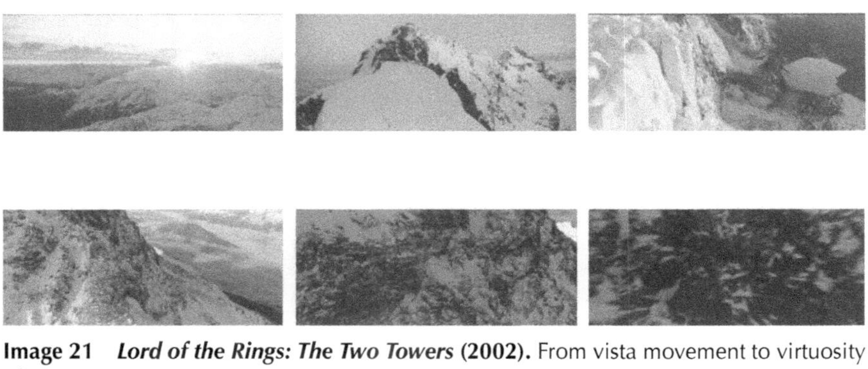

Image 21 *Lord of the Rings: The Two Towers* (2002). From vista movement to virtuosity of transport.

Hævnens nat (1916) which appears to move straight through the French door of Ann's room.[19] Especially the seemingly impossible traversals invite the viewer to appreciate the artifice of the move as when the camera penetrates the El Rancho-nightclub sign in *Citizen Kane* (Orson Welles, 1941) or when the camera appears to move through window bars at the end of Michelangelo Antonioni's *The Passenger* (Michelangelo Antonioni, 1975).

Wandering camera is a term that Seymour Chatman introduces (1985) and which Kenneth Johnson invokes in an article that explores wandering camera as a specific functional category of movement. The key characteristic of wandering camera is that the camera *detaches* itself from the most salient conveyors of story information. A good example of wandering camera is the camera movement in Martin Scorsese's *Taxi Driver* (1976) as Travis talks to Betsy on a pay phone. Initially the camera fulfills its characterological function showing us Travis taking in the words of rejection from Betsy but in the middle of the conversation, the camera—unmotivated by moving action—tracks right and stops on an empty corridor while we continue to hear Travis' side of the conversation offscreen.[20]

Another type of reflexive function is the *intertextual* function. Intertextual camera movements invoke a camera movement from another film or the camera movement conventions of another medium or genre (e.g., documentary, sports coverage). Intertextual camera movement can be invoked for parody as in *Bananas* (Woody Allen, 1971) or *Airplane!* (Jim Abrahams and the Zucker Brothers, 1980) which both mock a camera movement convention (push-in and swelling music to represent sudden realization and the arcing shot enclosing the embracing couple respectively) (Image 22).

The intertextual function is not always engaged for parody. In *Any Given Sunday* (Oliver Stone, 1999) the coverage of the NFL football games references camera mobility as it would have been staged in TV coverage (Steadicam shots moving *around* the pitch, cameras placed high up in the

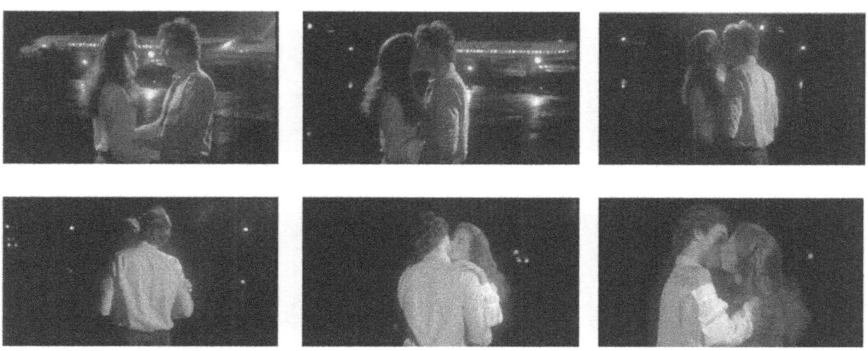

Image 22 The arcing-shot-around-the-embracing-couple is invoked, partially embraced yet certainly parodied at the end of *Airplane!* (1980).

stadium as well as overhead SkyCam shots) though without fully embracing it: The football games are also portrayed by means of an *on-the-field* Image Shaker-device which for obvious reasons cannot be used in TV coverage and which leaves the visual imprint of a stampeding herd. Used to seeing NFL football on the TV monitor the intertextual camera movements give the impression of an experience that is the *same but different.*

An intriguing type of reflexive camera movement is *parametric* movement. The term originates from Noël Burch's *Theory of Film Practice* (1973) and has been elaborated by David Bordwell in *Narration in the Fiction Film* (1985).[21] Parametric camera movement is an aesthetic strategy of moving the camera in a way that invites the viewer to be aware of (and to consider the implications of) stylistic patterning *across the course of the film.* In Agnés Varda's *Sans toit ni loi / Vagabond* (1985) a series of structurally similar follow shots are planted throughout the film. The shots share the following components:

a. The movements are parallel tracking shots from right to left.
b. At one time or another they show the main protagonist, Mona (Sandrine Bonnaire), vagabonding or simply walking in the same direction of the camera.
c. The sedate pace of the movement is roughly the same.
d. The same type of music accompanies the shots—Joanna Bruzdowicz's *Variations sur la vita.*
e. The moves come to rest on a static motif—an object from everyday life but often presented as an abstract form.[22]

The stylistic patterning of these camera movements—there are twelve in the course of the film—invite the spectator to diagram and analyze them as a part of a self-contained stylistic design. The patterning acquires a meaning that is largely independent of—or at least runs parallel to—the unfolding narrative chain of events. In order to get the spectator hooked onto this playful way of engaging with style, the film plants no less than four of the twelve moves within the first fifteen minutes of the film. In one sense these moves simply engage the spectator in the playful game of memory and recall involved in spotting similarities and variations of design: The first of the movements occurs approximately six minutes into the film and opens with the camera tilting up from a pedestrian crossing to a shot of Mona (Sandrine Bonnaire) approaching from the dunes near the sea; the follow shot is a parallel tracking shot that moves perpendicular to the background and retains approximately the same distance to Mona (Image 23).

The second of these parametric moves occurs only a few minutes later in the film but here the camera is angled at a diagonal to the background. In this second example Mona walks into the camera's field of view whereas it was the camera tilt that brought Mona into the field of view in the former example, the base of the dolly curves left at the end of the move and the actual camera pans left in the course of the movement, the shot scale fluctuation is different etc. (Image 24).

The parametric camera movements crop up at different times during the film but stitching them together, one finds that the end motif of the former camera movement is echoed in the opening motif of the subsequent camera movement. Consequently, the camera movements are tied together and form an entity in their own right (Image 25).

However, Varda does not simply stage a play of form: The implications of the aesthetic patterning *feed back* into the way in which we ascribe meaning to the behavior of the characters and our understanding of the narrative arc. The follow shots are invariably placed at intervals in the film where Mona is vagabonding. Mona is presented as restless, reluctant to settle down, and even more reluctant to take orders. Some of the characters that she encounters

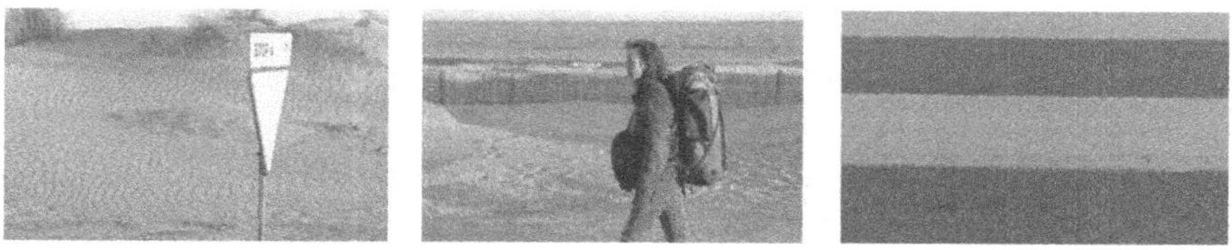

Image 23 ***Sans toit ni lois /Vagabond* (1985).** Parametric camera movement (read right to left).

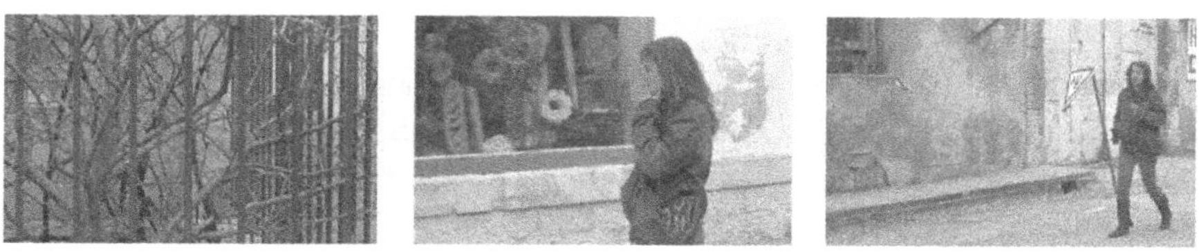

Image 24 ***Sans toit ni lois /Vagabond* (1985).** The second parametric camera movement (read right to left). *Note* the sign connecting the last shot in Image 23 and the first shot in Image 24.

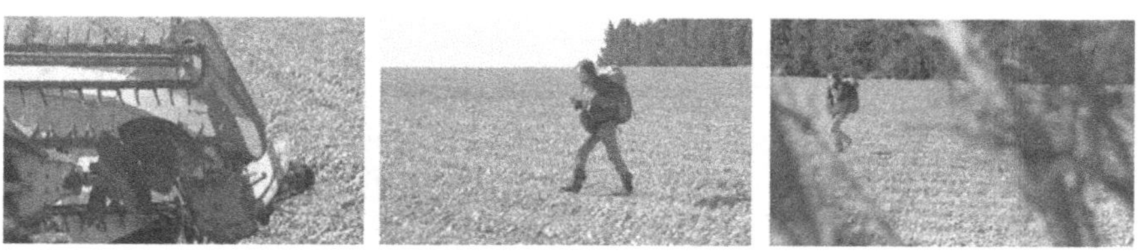

Image 25 ***Sans toit ni lois /Vagabond* (1985).** The third parametric camera movement (read right to left). *Note* the shrubbary connecting the last shot in Image 24 and the first shot in Image 25.

reveal to us that aspects of her way of life appeal to them or fascinate them. At times Mona herself seems to harbor romantic notions of walking the roads as a free and unattached individual without bosses looking over her shoulder. However, the directional patterning of the movements (of both Mona and camera) suggests that Varda wants to complicate the way in which we take in this outward appearance.

Western viewers are accustomed to read and write from left to right and rightward movement would be the logical way of portraying positive and progressive motion.[23] However, Varda reverts the movement, staging the shots so that camera and Mona move from right to left thereby suggesting that Mona's restless movement from place to place is not getting her anywhere but is in fact a kind of regression or as a farmer informs us: "She's not vagabonding, she is withering."[24]

In fact, as Mona is *not* walking the roads but is resting at a chateau together with a lover called David, *then* Varda invokes the pattern only to *reverse* it (Image 26). As David and Mona walk through the garden of the chateau, both they and the camera move *left to right*, the shrill violin sounds are replaced by serene music, and the shot does not come to rest on a street sign, harvester or fenced-in thorn bush but a much more organic motif: the leaves of a green plant (fig. 91–93). Invoking the aesthetic patterning and reverting it makes the viewer aware of a breach between the film and the filmmaker who takes a stance in relation to it.[25] In this particular film Varda navigates this breach in a way that allows the viewer to ascribe positive and negative values by means of the stylistic patterning. Hence the parametric camera movement in *Sans toit ni loi* also serves the valuative function described earlier.

Abstract

Camera moves may also suggest abstract ideas and concepts. The first definition of "abstract" given in *Webster's* is as follows: "conceived apart from concrete realities, specific objects, or actual instances: *an abstract idea*."[26] This is a useful definition because it allows us to distinguish abstract from

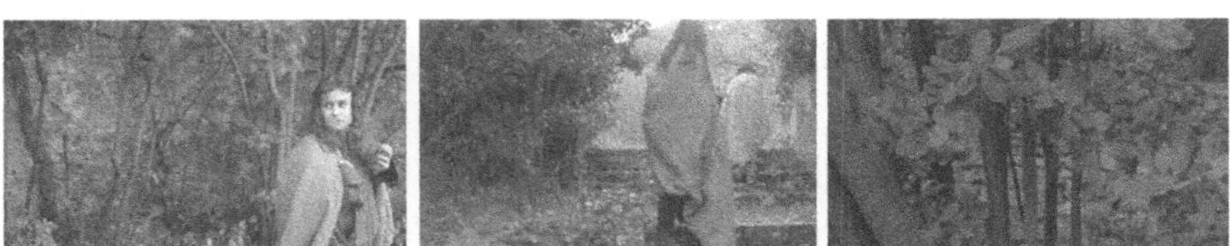

Image 26 ***Sans toit ni lois /Vagabond* (1985).** The film invokes the pattern in order to reverse it (read left to right). This time the tracking shot is also sliced in half: There is a cut between the first two shots.

other functions such as *evoke* which is still addressing the ability of camera movement to suggest *actual* subjective experience.

Two crane movements in *Far and Away* (1992) usefully illustrate the fine line at play. First the camera ascends through the roof and up toward the firmament as Joe Donelly dies, later the crane again ascends as we speculate that Joseph has died and his soul is leaving his body behind (Image 27)—only to descend rapidly again just before he wakes up again. If we take these camera movements to mediate the perspective of their souls, then they would be focalizing the experiential perspective of a metaphysical being. If on the other hand we take these camera movements to convey the idea of life leaving a body behind, then they suggest an abstract idea.

A particularly interesting example is the pendulous camera movement in the "dying-motorcyclist-scene" in *Der Himmel über Berlin* because it allows us to discuss various interpretations of the move in relation to the abstract function and various other functions launched previously.

This particular camera movement has been thoroughly researched by Richard Raskin who has brought up the example in interviews with Wim Wenders, director of photography Henri Alekan, Bruno Ganz, and Agnès Godard who operated the camera.[27] Furthermore, Raskin has himself written two articles on the movement that integrate interpretations suggested by his students, the explanations given by Wenders and Godard (Ganz and Alekan say little about the actual movement) as well as his own contribution to the debate which focuses on the verbal component of this and the surrounding shots.[28]

The scene in question takes place approximately one hour and fifteen minutes into the film and opens with a mobile shot: The camera moves forward smoothly yet quickly across a bridge. As the camera reaches the end of the bridge we see the scene of an accident: a smashed car with smoke rising from the hood, a motorcycle, and helmet lying on the ground as well as a number of bystanders. Near the right edge of the frame a motorcyclists rests awkwardly against the curb.

The initial pace and position of the camera suggests that it represents the point of view of a car but at the end of the move the camera curves to the right to focus attention on the motorcyclist in a manner impossible for an ordinary motor vehicle. There follows a cut revealing Damiel (Bruno Ganz) advancing

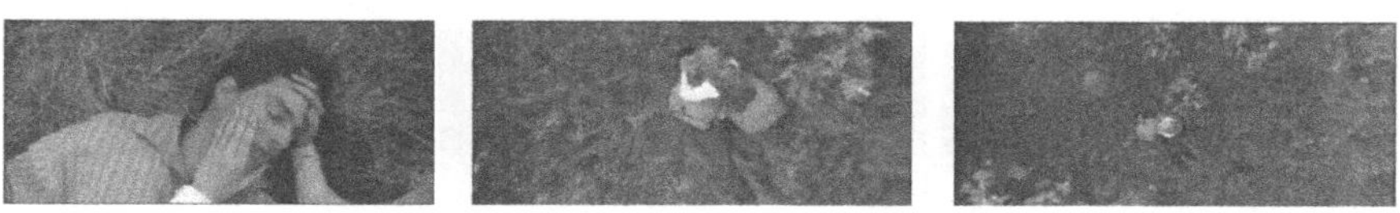

Image 27 ***Far and Away* (1992).** The camera ascends as Shannon believes Joseph to be dying.

slowly from roughly the position occupied by the camera at the end of the previous shot. Although there is no lead-in shot, the cut to Damiel suggests that the movement across the bridge embodied the point of view of Damiel who as an angel would be able to float across the bridge in the manner conveyed by the camera move. Substantiating this interpretation is that the inner voice of the motorcyclist—accessible only to angels—began to fade in on the soundtrack as the camera was crossing the bridge.

Damiel leans down toward the motorcyclist and places a hand on each side of his head. And then comes the slow pendulous camera move that Raskin embraces as "so original and striking in its form that it may be seen as a new cinematic figure, invented by Wenders to fulfill a specific need" (1999d, 159). First the camera tracks right, then left, then right and finally left again sometimes combined with small reframing pans.[29] Raskin has demonstrated how each of these movements correspond to a verbal change in the scene:

> When the camera first tracks right (2060 b), Damiel begins to recite the invocation for the dying man. When the camera then tracks left (2060 c), the dying man joins his voice to that of Damiel. When the camera once again tracks right (2060 d), the dying man alone recites the invocation. And when the camera tracks left for the final time (2060 e), the voice of the young man is heard, along with the dying man's invocation.[30]

In Raskin's articles and interviews a number of different abstract functions are ascribed to the camera move. To Wenders the movement visualizes the tenderness and pain involved as the angel Damiel is gradually taking over the dying man:

> I thought that going back and forth sort of showed more what Damiel was actually doing, in the way that he is - as the man is dying - that he's *taking him over*, so to speak. In a way, in a strange way, this "action" as well as the pain that came with it, were in that camera movement.
>
> [. . .] as it was somehow about a transition between life and death, it did translate something: not so much his p.o.v., more a mental attitude. Damiel's tenderness and his care for the man were in that back-and-forth movement.[31]

If the move functions to convey that Damiel is "taking him over" then the camera move conveys an abstract idea: for instance an angel accessing the mind of a dying man and helping him tune into an inner poem that already exists within himself but has to be unearthed. This is an interpretation suggested by Raskin but other interpretations of exactly what "taking him over" means also convey an abstract idea. If one takes the scene to mean that Damiel senses that the motorcyclist will actually die, then his actions are not aimed at rescuing the motorcyclist but at preparing him for the passage by redirecting his thoughts away from the fragmentary, despairing, disappointed, and regretful toward a tranquil and poetic formulation of what is significant in his life. Thus he will enter the realm of the dead with a serene mind.

In his interview with Agnès Godard, Raskin mentions another function of the camera move which some of his students had suggested: The camera move visualizes how the motorcyclist *oscillates* between life and death. Particularly as the action takes place at the end of a bridge—itself a transitional construction—this is a persuasive interpretation.[32] Godard takes a slightly different, but fundamentally similar, view of the shot. She takes the camera movement to be like "heartbeats that were slowing down" which would also mean that the camera "had to move in order not to be dead before the character. That's what I felt, in the rhythm, like a heartbeat that might stop at any moment."[33] Implicit in Godard's understanding of the shot is that if the camera stopped moving it would suggest that the character had died. In both cases the camera movement visualizes an abstract idea.

Just as significant are Godard's remarks about her initial fears about the swaying camera movements appearing reflexive: "I was afraid that in reframing the shot, in going after the young man, *we* were imposing—by the very fact that we went looking for him, that we were reframing the angel and the dying man—that *we* were pulling the strings of the story."[34] However, looking back at the pendulous camera movement Godard—and one can only agree with her—did not really fear that it would be reflexive in the sense of appealing to a level of engagement divergent from our engagement in the narrative. Camera movements can serve abstract functions without being reflexive if they are integrated within the world of the story.

To sum up, this article has presented five functions of camera movement in narrative cinema. There may be other functions but I believe these to be essential ones. Camera movements do not always fall neatly into one of the categories but the functions form the basis for combinatory variants. A way of analyzing camera movement is to picture a flow chart of layered functions. At one point a single function or a certain combination of functions will be engaged, sometimes to be replaced by another combination later in the same shot. Thus the taxonomy can be applied to both synchronic and diachronic analyses, that is, establishing what functions are at play in a particular segment of a shot and establishing how one set of layered functions weaves into another in the course of a shot.

Roman Jakobson's concept of the dominant is useful in this regard[35] although with regard to camera movement I have found that some functions typically form the platform for others, the primary ones being orientation and pacing. Hence I suggest that one studies a camera movement from the top-down so to speak—asking first in what way it orients the viewer spatially, then how it is temporally articulated before moving on to consider the other functions.

It is important to stress that a functional analysis of camera movement does not represent a fully fleshed analysis of the relevant shots and scenes in which the camera movement takes place. Camera movement is one of many devices in the polyphony of cinematic expression. However, a better grasp of the functions of camera movement leads to a better understanding of the full meaning of a shot.

This is one of the staggering tasks that lie ahead but these five functions can hopefully assist many analytic and interpretative procedures or enterprises. It may assist analyses of individual camera movements within a given film but one can also use it as a framework for discussing authorial camera movement or genre-specific camera movement. For instance what functions do this particular director or genre incline toward? One can also ask what functions are engaged by films at a particular point in time. Do contemporary blockbusters explore abstract functions more than the Westerns did in the 1940s? The latter question resides within the realm of historical poetics.

A lot of work still remains to be done on style generally and on camera movement in particular. Even after several years of researching camera movement they can still baffle and surprise a trained eye and prompt me to revisit a specific movement again and again. One must look long and hard to capture the elusive effects of camera movement.

NOTES

1. The following is an abridged and revised version of a chapter in my PhD dissertation *Camera Movement in Narrative Cinema—Towards a Taxonomy Functions* (Aarhus University, 2007).

2. Nielsen, *Camera Movement*, 87–218.

3. Allan Dwan in Peter Bogdanovich, *Who The Devil Made It?* (New York: Alfred A. Knopf, 1997), 97.

4. Vilmos Zsigmond to the camera in Greg Carson (dir.), *Vilmos Zsigmond Flashes 'The Long Goodbye.'* MGM Home Entertainment (2002).

5. David Bordwell, "Camera Movement and Cinematic Space." *Cine-Tracts* 1, 2 (Summer 1977): 19–26.

6. Garrett Brown, "The Moving Camera Part I." *Zerb* 58 (Autumn 2003). Also available on www.garrettcam.com.

7. Miller quoted in David Bordwell and Kristin Thompson, *Film Art: An Introduction,* 7th ed. (N.Y.: McGraw-Hill, 2004): 269.

8. For a discussion of surface/depth-articulation in Japanese film, see Noël Burch, *To the Distant Observer* (London: Scholar Press 1979), 127–132.

9. To the camera in Arnold Glassman, Todd McCarthy and Stuart Samuels (dir.), *Visions of Light* AFI/NHK (1992).

10. Burch, *Distant Observer*, 225 and Lutz Bacher, *Max Ophuls in the Hollywood Studios* (New Brunswick: Rutgers UP, 1996): 5.

11. Siegfried Kracauer, *From Caligari to Hitler* (Princeton: Princeton UP, 1947): 104–5. Kracauer's description is not altogether precise as it suggests that the camera moves from A (scarecrow) to B (the couple in the window). In fact, the camera only tilts up ever so slightly keeping the scarecrow in the lower right foreground of the frame while the window pops into view in the upper left corner of the frame. Kracauer also fails to mention the insinuations of rape. At least the print of the film that I have seen does not present a silhouetted embrace. In fact, one cannot clearly see the silhouetted actions except for what looks like two shadows moving *across* the window as if one was pursued by another.

12. For an analysis of the opening camera movements of *Sunnyside Up* (1929) and *Gravity* (2013), see Jakob Isak Nielsen, "Sunnyside Up." *16:9* no. 43 (September 2011) and Jakob Isak Nielsen, "Blockbusteren som eksperimentalfilm." *16:9* no. 54 (February 2014). www.16–9.dk. Accessed 1 March 2016.

13. Paul Thomas Anderson, "DVD-commentary track." *Hard Eight/Sidney*. Columbia/TriStar (1999).

14. This breaking down of the components involved in cinematic rhythm has its origin in David Bordwell, *The Films of Carl Theodor Dreyer* (Berkeley: University of California Press, 1981): 140–3.

15. See for instance David Bordwell, *Figures Traced in Light* (Berkeley, L.A.: University of California Press, 2005).

16. Some experimental studies indicate that choice of camera support in itself registers differently in the spectator's motor cortex and thus causes different forms of involvement, see Katrin Heimann, Maria Alessandra Umiltà, Michele Guerra, and Vittorio Gallese, "Moving Mirrors: A High-density EEG Study Investigating the Effect of Camera Movements on Motor Cortex Activation during Action Observation," *Journal of Cognitive Neuroscience* 26(9) (2014): 2087–2101. An empirical study—not unlike the Kuleshov experiment—deserves to be made to determine whether or not a push-in on an *uninflected shot* (a character whose mimicry and gestures are minimized) will suggest psychological activity on the part of the character whom we are witnessing.

17. So established is this convention that Woody Allen can effectively mock it in *Bananas* (1971). The joke *depends* on the viewer's belief in the commentative function of the push-in for Allen to be able to deflate the convention.

18. See Lutz Bacher, *Max Ophuls's Universal–International Films: The Impact of Production Circumstances on a Visual Style* (PhD dissertation, Wayne State University, 1984) and Bacher, *Hollywood Studios*.

19. Nielsen, *Camera Movement*, 264–270.

20. For an analysis of this camera movement, see Jakob Isak Nielsen, "Antonioni Pirouette," *16:9* no. 39 (November 2010). Accessed 1 March. www.16–9.dk

21. Noël Burch, *Theory of Film Practice* (London: Secker & Warburg, 1973) and David Bordwell, *Narration in the Fiction Film* (London: Routledge, 1985).

22. I am grateful to Ben Singer for drawing my attention to these camera movements.

23. There is statistical evidence to suggest that rightward movements outnumber leftward movements in Hollywood films, see Brian O'Leary, "Hollywood Camera Movements and the Films of Howard Hawks: a Functional Semiotic Approach." *New Review of Film and Television Studies* 1, 1 (November 2003): 21–22.

24. Another implicit meaning that one can infer from Varda's directional strategy is that Mona may be taken to be eternally walking toward home. Just as an early camera movement in Mike Nichols' *The Graduate* (1967) prepares us for a going-home-experience by moving laterally with Benjamin (Dustin Hoffman) from right to left as he steps unto a rolling escalator, the leftward follow shots may have similar implications in *Sans Toit ni Loi* only Mona has no home to return to.

25. Alexandre Astruc, "The Birth of a New Avant-Garde: La Caméra-Stylo" *The New Wave*, ed. Peter Graham (Garden City, N.Y.: Doubleday and Company Inc., 1968). Originally published as "Naissance d'une nouvelle avant-garde, la caméra-stylo," in *L'Ecran Français* no. 140 (March 30, 1948).

26. *Webster's Encyclopedic Unabridged Dictionary of the English Language* (Rev. edition, 1994).

27. Richard Raskin, "It's *Images* You Can Trust Less and Less," *p.o.v.* no. 8 (December 1999): 5–20; "If There Is Such a Thing as Real Angels," *p.o.v.* no. 8 (December 1999): 21–37; "Bringing Images to Life," *p.o.v.* no. 8 (December 1999): 38–46.

28. Richard Raskin, "Camera Movement in Wings of Desire." *p.o.v.* no. 4 (December, 1997): 79–100. Richard Raskin, "Camera Movement in the Dying Man Scene in Wings of Desire." *p.o.v.* no. 8 (December 1999): 157–70.

29. The various segments of the camera movement are illustrated at http://pov.imv.au.dk/Issue_04/section_1/artc5A.html#RTFToC4. Accessed 1 August 2016.

30. Raskin, *Dying Man Scene*, 165–166. Raskin cites the screenplay in which the inner voice is referred to as "The invocation of the world."

31. Raskin, *Images You Can Trust Less and Less*, 7. Also cited in Raskin, *Camera Movement*, 91.

32. Raskin, *Bringing Images to Life*, 43.

33. Raskin, *Bringing Images to Life*, 40. While Wenders also argues that the camera movement has to do with the transition between life and death, he does not exactly describe the camera movement as visualizing the way in which the man drifts between life and death.

34. Raskin, *Bringing Images to Life*, 44.

35. Roman Jakobson, "Linguistics and Poetics." In *Style in Language* (1960), edited by Thomas Sebeok. Reprinted in *Modern Criticism and Theory* (1988), edited by David Lodge, 32–56. London & N.Y.: Longman.

BIBLIOGRAPHY

Anderson, Paul Thomas. 1999. DVD-commentary track. *Hard Eight/Sidney*. Columbia/TriStar.

Astruc, Alexandre. 1968 [1948]. "The Birth of a New Avant-Garde: La Caméra-Stylo" *The New Wave*, ed. Peter Graham (Garden City, N.Y.: Doubleday and Company

Inc., 1968). Originally published as "Naissance d'une nouvelle avant-garde, la caméra-stylo," in L'*Ecran Français* no. 140 (March 30, 1948).

Bacher, Lutz. 1984. *Max Ophuls's Universal–International Films: The Impact of Production Circumstances on a Visual Style*. Ph.d. dissertation, Wayne State University.

———. 1996. *Max Ophuls in the Hollywood Studios*. New Brunswick: Rutgers UP.

Bogdanovich, Peter (1997). *Who the Devil Made It: Conversations With Legendary Film Directors*. N.Y.: Ballantine Books.

Bordwell, David. 1977. "Camera Movement and Cinematic Space." *Cine-Tracts* 1, 2 (Summer): 19–26. Reprinted in Ron Burnett, ed., *Explorations in Film Theory: Selected Essays from Ciné-tracts* (Bloomington: Indiana UP).

———. 1981. *The Films of Carl Theodor Dreyer*. Berkeley: Univ. of California Press.

———. 1985. *Narration in the Fiction Film*. London: Routledge.

Bordwell, David & Thompson, Kristin. 2004. *Film Art: An Introduction* (7th ed.). N.Y.: McGraw-Hill.

Bordwell, David. 2005. *Figures Traced in Light*. Berkeley, L.A. and London: University of California Press.

———. 2006. *The Way Hollywood Tells It*. Berkeley: University of California Press.

Brown, Garrett. 2003. "The Moving Camera Part I." *Zerb* 58 (Autumn). Also available on www.garrettcam.com.

Burch, Noël. 1973. *Theory of Film Practice*. London: Secker & Warburg.

Burch, Noël. 1979. *To the Distant Observer*. London: Scholar Press.

Carson, Greg, dir. (2002). *Vilmos Zsigmond Flashes 'The Long Goodbye.'* MGM Home Entertainment.

Glassman, Arnold; McCarthy, Todd; Samuels, Stuart. 1992. *Visions of Light*. AFI/NHK. A Documentary on Cinematography.

Heimann, Katrin, Maria Alessandra Umiltà, Michele Guerra, and Vittorio Gallese. 2014. "Moving Mirrors: A High-density EEG Study Investigating the Effect of Camera Movements on Motor Cortex Activation during Action Observation," *Journal of Cognitive Neuroscience* 26(9): 2087–2101.

Jakobson, Roman. 1960. "Linguistics and Poetics." In *Style in Language*, edited by Thomas Sebeok. Reprinted in *Modern Criticism and Theory* (1988), edited by David Lodge, 32–56. London & N.Y.: Longman.

Kracauer, Siegfried. 1947. *From Caligari to Hitler*. Princeton: Princeton UP.

Murnau, F. W. 1928. "Films of the Future." *McCall's Magazine* (September): 27, 90.

Nielsen, Jakob Isak. 2007. *Camera Movement in Narrative Cinema: Towards a Taxonomy of Functions*. PhD dissertation, Aarhus University.

Nielsen, Jakob Isak. 2010. "Antonioni Pirouette." In *16:9*, no. 39 (November). http://www.16–9.dk/2010–11/side11_inenglish.htm

———. 2011. "Sunnyside Up." In *16:9*, no. 43 (September). http://www.16–9.dk/2011–09/side07_anatomi.htm

———. 2014. "Blockbusteren som eksperimentalfilm." In *16:9* no. 54 (February). http://www.16–9.dk/2014–02/side06_feature3.htm. Also available at http://www.16–9.dk/2014/02/blockbusteren-som-eksperimentalfilm/

O'Leary, Brian. 2003. "Hollywood Camera Movements and the Films ofHoward Hawks: a Functional Semiotic Approach." *New Review of Film and Television Studies* 1, 1 (November): 7–30.

Raskin, Richard. 1997. "Camera Movement in Wings of Desire." *p.o.v.* no. 4 (December): 79–100.

———. 1999a. 'It's Images You Can Trust Less and Less', *p.o.v.* no. 8 (December): 5–20. An Interview with Wim Wenders on *Wings of Desire.*

———. 1999b. "If There Is Such a Thing as Real Angels," *p.o.v.* no. 8 (December): 21–37. An Interview with Henri Alekan on *Wings of Desire.*

———. 1999c. "Bringing Images to Life," *p.o.v.* no. 8 (December): 38–46. An Interview with Agnès Godard on *Wings of Desire.*

———. 1999d. "Camera Movement in the Dying Man Scene in Wings of Desire." *p.o.v.* no. 8 (December): 157–70.

Salt, Barry. 1992. *Film Style and Technology: History and Analysis* (2. ed.). London: Starword.

Webster's Encyclopedic Unabridged Dictionary of the English Language (Rev. edition, 1994).

Chapter 3

À Travers

The Cinematography of Depth in Japan

Daisuke Miyao

BEYOND SUPERFLAT

> Society, customs, art, culture: all are extremely two-dimensional. It is particularly apparent in the arts that this sensibility has been flowing steadily beneath the surface of Japanese history. . . . This book hopes to reconsider "super flatness," the sensibility that has contributed to and continues to contribute to the construction of Japanese culture, as a worldview, and show that it is an original concept that links the past with the present and the future. During the modern period, as Japan has been Westernized, how has this "super flat" sensibility metamorphosed?. . . . "Super flatness" is an original concept of Japanese who have been completely Westernized.[1]

In April 2000, the Japanese visual artist Murakami Takashi curated an exhibition titled *Superflat*. By "Superflat," Murakami refers to various flattened forms in Japanese graphic art, animation, pop culture and fine arts. Moreover, in "The Super Flat Manifesto," the introductory essay to the exhibition catalog, Murakami tries to create a direct lineage between art from Japan's premodern period and contemporary art, including anime and manga. Murakami's "superflat" essentializes a Japanese aesthetics "sensibility" that runs through history as if the period of modernity and modernization had never existed in Japan. As a first example of superflat, the catalog juxtaposes a series of stills of the explosion of a planet from *Ginga tetsudo 999* (*Galaxy Express 999*, 1979) with a famous woodblock print from Hokusai's *Fugaku sanjurokkei* (*Thirty Six Views of Mt. Fuji*, c. 1831). Thomas Lamarre describes, "Both images flatten the relationship between different planes (foreground, background, and middle grounds), with zigzagging, arching, sweeping lines that encourage the eye to wander over the surface of the image, restlessly scanning it."[2] As Lamarre points out, Murakami makes the connection "entirely in terms of the structural composition of the image," which apparently lacks depth.[3] The movement of "the eye" is, according to Murakami, strictly limited to the flat surface. Lamarre argues that the notion of superflat "hinges on establishing a fundamental difference between Japanese traditions of flat or planar composition on the one hand and Western traditions of one-point, linear, or geometric perspective on the other."[4] The culturally essentialist positioning of superflat theory avoids dealing with questions about modernity and modernization of Japan, which have had a complicated relation to Western technology. Discussing the cinematography of depth from a historical perspective, this essay introduces one aspect of such complications in visual art and technology in Japan. Like Murakami's superflat theory, my focus is also on the movement of the eyes. However, I will not argue that the movement is limited to the surface no matter how flat the images of Japanese art and cinema look compared to those in Western counterparts. Even in those flat-looking images, cinematography invites the eyes of the viewers to be more unstable, mobile, and ephemeral not only on the surface level but also on the depth of fields.

THE LION KING CONTROVERSY AND THE *À TRAVERS* CINEMA

When Disney's animated musical film *The Lion King* (Roger Allers and Rob Minkoff, 1994) was released, the *San Francisco Chronicle* reported a protest from those who said the film borrowed substantially from *Jungle Emperor* (*Janguru taitei*), a manga by Tezuka Osamu (1928–1989) originally published in 1950–1966; an animated TV series (1965–1966) based on the manga. *Jungle Emperor* was also broadcasted in the United States with the title, *Kimba, the White Lion* (1966). The protest was heightened because Disney did not credit Tezuka despite such similarities, even though Disney might not have been obligated to do so.

Curiously, the response from Tezuka Productions in Tokyo was not critical of Disney. Matsutani Takayuki, president of Tezuka Productions said, according to the report, "If Disney took hints from 'The Jungle Emperor,'" our founder the late Osamu Tezuka, would be very pleased by it. . . . Rather than filing a claim, we would be very happy to know that Disney people saw Tezuka's work. On the whole, we think 'Lion King' is absolutely different from 'Jungle Emperor' and is Disney's original work."

It is certainly possible to point out that Tezuka was a big fan of Disney animation and he must have felt honored if Disney really imitated his work. It is said that Tezuka watched *Bambi* (David Hand, James Algar, Bill Roberts, Norman Wright, Samuel Armstrong, Paul Satterfield, Graham Heid, 1942) eighty times and *Snow White and Seven Dwarfs* (David Hand, Larry Morey, Wilfred Jackson, Ben Sharpsteen, Perce Pearce, William Cottrell, 1938) fifty times in theaters when these films were released in Japan after World War II ended. He even published manga versions of *Bambi* and *Pinocchio* in Japan in 1951 and 1952 respectively.

Matsutani's statement that *The Lion King* is "absolutely different" from *Jungle Emperor* becomes more interesting when we look at *Jungle Emperor Leo* (*Janguru taitei*, Takeuchi Yoshio, 1997), a new animated feature that Tezuka Productions released three years after the controversy around *The Lion King*. Curiously, the 1997 version does not begin from the very beginning of the original manga but from the midst of its story, when Lune and Lukio, the babies of Leo, the lion and the respected king of the African jungle, are born. Seemingly, this opening follows *The Lion King*, which starts when Mufasa's son, Simba, is born.

While the beginnings of the stories are identical, the images that we see on the screen look quite different. The openings of *The Lion King* and *Jungle Emperor Leo* indicate "different" ways of representing the world in the cinematic, or animated cinematic form. The major focus of the opening of *The Lion King* is the emphasis on depth. After the first extreme long shot of the daybreak, there is a series of medium shots of animals (a rhino, horned deer, meerkats, a cheetah, and pelicans). Following a medium shot of a pelican starting to fly, there are a series of extreme long shots. These extreme long shots are typical examples of multi-plane camera that preceded 3-D computer generated animation that simulates the visual qualities of three-dimensional space as it is photographed.[5] A flock of pelicans fly in front of giant falls. A herd of elephants walk in the mist in front of a high mountain. A group of flamingos fly over the river when more birds pass across the frame closer to the camera. A herd of horned deer jump across the frame at back and in front. A baby giraffe appears from behind a hill, accompanied by its mother, approaches the camera, makes a slight left turn, and looks over the hill. What we see with the baby giraffe, which stays in the left front corner of the frame, is a number of different animals walking in the vast field at a great distance. In the same shot, the two giraffes start to run and join the other animals at the far back. In the next shot, a line of bugs carry green leaves on a tree branch in front while a herd of zebras run on the ground behind the tree. A shot of blue birds and elephants running toward the camera follows. Eventually, all these animals gather under a gigantic rock. Following the movement of a flying bird, the camera swiftly approaches the rock where Mufasa stands. What is important in this scene is photographic realism, or recreation of liner perspective, within the computer-enhanced animation. Visual verisimilitude is maintained clearly by linear perspective. Thomas Looser argues:

> In classic Western perspectival space, everything is unified and hierarchized by the single vanishing point; space itself is accordingly homogenous, and everything finds its proper place within that space, including the spectator (drawn into the picture plane via the vanishing point), in accordance with fixed mathematical laws of relation. This yields stable positions of near versus far, subject versus object, and, supposedly, the position and identity of the viewing subject. Again, in a depth model, that now seems almost three-dimensional (but qualitatively homogenous).[6]

Compared to the opening of *The Lion King*, that of *Jungle Emperor Leo* does not look fully three-dimensional. The movements of animals look flat. Many of their movements are horizontal or vertical, but not between the front and the back. Leo's approach to his home, a ruin, is depicted as a POV shot, and we do not see Leo physically come close to the place. Cel animation uses meticulously painted outline drawings on thin, transparent sheets of celluloid. Cel animation, in which pictures are drawn directly onto celluloid "cels," has been the most common technique of animation. Cel animation began in the United States in 1913, and from around 1929, the same technique was used in Japan. Images are logically displayed in two-dimensional space, or "superflat" space, if you like.

Yet, what I want to stress here is not the flatness of images but a specific expression of depth in the anime's cinematography. The expression also invites awkward but interesting movements of the eyes. It does not faithfully follow Western sense of linear perspective but creates rather jerkey movements among different layers. When cel animation integrates depth, it tends to be represented between layers of composited images. Leo's movements occur in front of the background, including the waterfall and the sky, that is drawn in multiple layers and photographed on equipment called animation stand, a simple apparatus for stacking celluloid sheets. When he jumps from the top of a waterfall to the bottom, Leo becomes small then big then small again in an attempt to follow the rules of linear perspective, yet the sizes of the clouds in the sky and the waterfall in the background do not change proportionally. This makes Leo's potentially three-dimensional movement look awkward. With 3-D computer generated spaces, when the point of view changes, proportions of objects change accordingly. In contrast, in cel animation, because all the images are drawn and painted by hands, it is difficult to recreate such proportional changes of objects. As a result, when characters move between front and back, we cannot help noticing the multiple layers in which the images are drawn. Such use of layers is clear in the opening scene when an elephant and a giraffe raise Lune and Lukio high in the air using their long trunk and long neck. The same technique is used. All the other animals on the ground are shown in a flat and static drawing in the background layer. Only Lune, Lukio, the elephant and the giraffe move in the front layer. Even though the sizes of Lune and Lukio change as they are lifted from the ground, the sizes of the other animals do not change proportionally. The camera perspective moves up as the elephant's trunk and the giraffe's neck are raised. The combination of these movements creates a strange mismatch of sizes in the supposedly three-dimensional movement. As a result, the eyes of the viewers cannot help moving back and forth, in depth, on a flat-looking surface of a cel animation.

This movement between back and forth comes close to the notion of *à travers* that art historian Mabuchi Akiko suggests when she discusses Claude Monet's paintings.[7] According to Mabuchi, Monet used a composition that emphasized the contrast between the front and the back in order to express the sense of vision that transits or moves from one end of a plane to another.[8] Monet achieved the *à travers* theory arguably by way of the use of depth in *ukiyo-e* woodblock prints by the Edo artists. On surface, woodblock prints of by Hiroshige or Hokusai flatten the hierarchal relationship of depth between the front and the back, as Murakami insists in "The Superflat Manifesto." However, more importantly, many of the Edo artists knew and understood Western geometric perspective perfectly and used it "as one among many modes of spatial organization that could be played with, and even layered over other kinds of space," as Looser claims.[9] Some of Hiroshige's images, according to Looser, often play with two or more viewpoints within the same image, inviting the viewers not so much to reject depth as to layer viewpoints. Referring to Looser, Thomas Lamarre also concludes that in these woodblock prints "layering takes priority over positioning, which opens the viewpoint to multiplication and distortion, to play and divergence."[10] Such multiplication, distortion, play and divergence of the movements of the eyes are enhanced by the apparatus of animation stand and the cinematography for the opening of *Jungle Emperor Leo*. I want to name this the *à travers* cinema.

Thus, if we can make a connection between woodblock prints and the style of cel animation, it should not be in terms of the ahistorical theory of superflat but by way of historical contextualization of how the Edo artists incorporated Western linear perspective and the notion of depth in the two-dimensional image. Also, the technology of anime, the animation stand and its cinematography, played a significant role to create the *à travers* effect.

CINEMATOGRAPHER MIMURA AKIRA AND THE "COMPOSITIONAL LIGHTING"

Discussions on the cinematography of depth first appeared in Japan in the 1930s. They were over the films directed by young filmmaker Yamanaka Sadao (1909–1938). Examining Yamanaka's second film *Money Trickles*

Down (*Koban shigure*, 1932), film critic Aikawa Kusuhiko named Yamanaka's technique of constructing images the "composition of depth" (*tate no kozu*). Aikawa states, "Director Yamanaka Sadao constantly constructs relationships between people or between people and objects in depth—a perspective and three-dimensional manner. He wonderfully and effectively uses this 'composition of depth' in order to express character psychology, to formulate certain atmospheres of scenes, and so forth."[11] As an example, Aikawa discusses a scene of the heroine O-Yone tragically separating from her lover. She sits down at a bar that she manages and laments. Aikawa writes, "Yamanaka Sadao depicts depressed O-Yone through a rope curtain of the bar. In this case, the rope curtain slowly sways from right to left as if it reflects O-Yone's uneasy feeling. The movement of the rope curtain at front and the movement of O-Yone's mind are adequately in harmony. We are witnessing a intelligent expression of human psychology by the 'composition of depth.'"[12]

Emphasizing the contrast and the connection between the front and the back within images, critics immediately connected Yamanaka's "composition of depth" to the work of the Edo artists, including Hiroshige. Critic Sugiyama Heiichi stated in 1941 that Yamanaka "invented the 'composition of depth'. . . inspired by Hiroshige."[13] Sugiyama defined the "composition of depth" as "the term that describes the method of capturing reality by combining the front and the back of an image."[14] Thus, Yamanaka's cinematic composition of depth was first examined in conjunction with the woodblock prints of the Edo period. Here, it is clear that critics did not consider Hiroshige's work to be simply flat. Instead, their emphasis was on the *à travers* effect, the movement of the eyes of the viewers between the heroine at back and the rope curtain at front.

The discussion on the composition of depth became more complicated when a Japanese cinematographer returned from Hollywood and started working at the same studio where Yamanaka directed his films. The cinematographer was aware of the experiments of deep-focus cinematography in Hollywood. Especially when they collaborated for a film in 1937, the ideas on depth in cinema became more complicated and global. It was no longer a simple dichotomy between the culturally essential superflat sensibility of Japan and the photographic realism of three-dimensionality of Hollywood and West.

Mimura Akira (1901–1985) was the first Japanese cinematographer who obtained membership in Hollywood's cinematographers' union (International Photographers of IATSE, Local 659). After studying at the New York Institute of Photography with Carl Gregory, D. W. Griffith's company man, Mimura started his career as an assistant to the acclaimed cinematographer George Barnes in Hollywood in the late 1920s. Gregg Toland was an assistant to Barnes at the same time. One of the first works of Mimura with Barnes and

Toland was *Trespasser* (Edmond Goulding, 1928), a star vehicle for Gloria Swanson.[15] After Barnes left the Goldwyn studio, Mimura, until November 1933, worked for Toland.

Mimura was brought back to Japan by the Photo Chemical Laboratory studio (future Toho) to become the ace cinematographer for the new company. Mimura's biggest challenge was the limited material conditions in Japanese cinema. Mimura constantly showed his awareness to the innovative work of his rival, Gregg Toland, in Hollywood. In 1948, Mimura wrote on Toland's work in *Best Years of Our Lives* (William Wyler, 1946), "The most notable issue in the cinematography of this film is crafty lighting. Throughout the film, he [Toland] creates vivid and sharp images. Appropriate to its content, the tone of the film appears as that of a documentary film."[16] Mimura was thus inspired by Toland's work that explored compositions of depth of field in order to give stories harsh, brutal moods, instead of simply clarifying the story with carefully controlled focus on images. Despite physically locating their films in different spaces, Toland and Mimura were in tune in pursuing "the illusion of presence" on the screen.[17]

When Mimura came back to Japan in 1934, Japanese critics and cinematographers were in the midst of a boom of praising the low-key cinematography of Hollywood, the "diverse expressivity in terms of gradation" which was different from the dark tones or strong contrast between light and shadow that existed even before the advent of supersensitive panchromatic film.[18] They identified the advent of Supersensitive Panchromatic Type Two Motion Picture Negative film by Eastman Kodak Company, which was announced on February 5, 1931, as the key moment in the birth of the low-key tendency in Hollywood lighting. They understood that ample lights and a variety of lamps on the sets would be needed in order to be perfectly balanced to achieve the low-key tones. However, cinematographers and critics in Japan were aware that such conditions were unimaginable given the reality of material and financial limitations in Japan. Soyama Naomori, who worked at the J.O. studio in Kyoto, wrote, "People in this country need to learn the American methods [of cinematography]. However, financially it is nearly impossible to follow them. In order to realize ideas and motifs only with available lighting and shallow lenses, we need to sacrifice the final process that American methods require."[19]

As I have discussed elsewhere, Japanese cinematographers and critics needed to find a way to deal with this gap between the ideal cinematography and the material limitation.[20] *Kage no bigaku*, or the aesthetics of shadow, was widely adopted among cinematographers and critics under such conditions in order to justify the gradation-less shadows that Japanese cinematographers could only achieve in their efforts of low-key lighting. Midorikawa Michio, the head of the Nipponese Society for Cinematographers (Nihon

Eiga Kameraman Kyokai), claimed, "We should observe the beauty of shadows, which appears gracefully in the harmony of [Japanese] architecture and lights."[21] The aesthetics of shadow was born as a strategic bending of Hollywood's trend of low-key lighting in the name of Japanese traditional culture in order to embrace the limited material conditions.

Gregg Toland's deep-focus cinematography of *Citizen Kane* (Orson Welles, 1941) was realized in the ample material and financial conditions in Hollywood. The keynote for Toland was realism. Toland discussed with Orson Welles that in the film, "the picture should be brought to the screen in such a way that the audience would feel it was looking at reality, rather than merely at a movie."[22] "The first step" that Toland and Welles took "was in designing sets which would in themselves strike the desired note of reality."[23] The result was that the majority of their sets for *Citizen Kane* had actual ceilings. Conventionally in Hollywood, interior light effects come from angles that "would be definitely impossible in an actual, ceilinged room" because "the light is projected from spotlighting units perched high on the lamp-rails paralleling the sets."[24] Since the sets were "ceilinged," according to Toland, the sets for *Citizen Kane* weren't paralleled for overhead lighting and almost everything in the picture was to be lighted from the floor.[25]

In addition to the production design and the lighting angles, Toland and Welles tried to "obtain the definition and depth" of the human eye in real life and not to require audiences to see things on the screen "with a single point of perfect focus, and everything falling off with greater or less rapidity in front of and behind this particular point."[26] To do so, Toland and Welles increased the illumination level by adopting the hard light of arc broad lamps, which were generally used in Technicolor, in addition to using Super xx film "with a super-speed emulsion."[27] As a result, according to Toland, "While in conventional practice, even with coated lenses, most normal interior scenes are filmed at maximum aperture or close to it—say within the range between f:2.3 and f:2.8, with an occasional drop to an aperture of f:3.5 sufficiently out of the ordinary to cause comment—we photographed nearly all of our interior scenes at apertures not greater than f:8—and often smaller. Some scenes were filmed at f:11, and one even at f:16!"[28]

According to Patrick Ogle, the introduction of Eastman Plus-x Panchromatic Negative in late October 1938 and Super xx just two weeks later with their greater film speed "developed initially for newsreel work and other specialized and realistic filming done under difficult lighting conditions" could have "reversed" the "tonal softness" that characterized Hollywood cinema in the 1930s.[29] A few cinematographers, such as Victor Milner, became immediately aware of the possibilities of the increased crispness and depth of field. Milner claimed, "It [supersensitive panchromatic film] makes it possible for us to run the scale between extremely soft, naturalesque low-level lightings (50 foot candles or less), shot with full lens apertures, to the opposite extreme of higher level illumination (perhaps as high as 200 foot candles or more) exposed at greatly reduced apertures for a new and greater depth and crispness."[30] Yet, according to Ogle, "Conservative by nature and distressed by the increased contrastiness of the new general-use Plus-x emulsion," the greatest number of cinematographers (Milner included) employed Plus-x and Super xx "with deliberate underdevelopment, a procedure that gave them lowered contrast" and preserved the soft tonality.[31] Toland's choice used the potentiality of the fast film stock but went against the general soft tendency of Hollywood cinematography in order to pushed forward "the illusion of presence."

Toland's obsession with realism was bound for admiration in the discourse of the aesthetics of shadow, which prioritized the documentary quality of darkness to the glamorous gradation. Still, it was nearly impossible to replicate the expensive production of a film like *Citizen Kane* in Japan. In August 1941, Mimura wrote in his article praising the cinematography in *Citizen Kane*, "I cannot help expecting the appearance of Japanese-made products that can compete with American film technologies, including supersensitive films, coated lenses, and high-intensity carbon lamps."[32] Mimura's aspiration sounded desperate because he was attempting a similar thing—"the illusion of presence" without using either modern lamps or the most-advanced film stock.

In September 1938, Shimazaki Kiyohiko of the Japanese Association of Film Technology (Nihon Eiga Gijutsu Kyokai) criticized the work of Mimura, who, according to Shimazaki, "learned nice techniques directly from the world of American cinematography" and "ended up with a simple imitation."[33] Shimazaki might have been correct when he pointed out Mimura's link to Hollywood, but what he overlooked was that Mimura was not simply imitating Hollywood but exploring ways to achieve "three-dimensionality" on the screen, which would go along with the "compositional lighting" that Shimazaki insisted as a Japanese response to the Hollywood cinematography. Shimazaki claimed, "Truly cinematic compositions of images are not yet established in Japanese cinema," and he held up the importance of "composition with all possible styles, techniques, and technologies as meticulous and organic as possible, not in two- but in realistic three-dimensionality."[34] Shimazaki argued, "By composition, I do not only mean two-dimensional arrangements of such elements as straight lines, winding lines, and so on, but nonflat ones with lighting and tones. In addition, all of those elements need to move, transform, and have three-dimensionality."[35]

Mimura's work in *Humanity and Paper Balloons* (*Ninjo kamifusen*, Yamanaka Sadao, 1937), is a good example of his conception of depth cinematography. His utilization of the "compositional lighting" distinguishes his depth cinematography from the photographic realism, or the emphasis on liner perspective, of Toland's work in *Citizen Kane*. It is also possible to argue that the *à traverse* effect is realized in Mimura's work. But such an observation should not come from a transhistorical, culturally essentialist connection between the 1937 film and Hiroshige's woodblock prints of the nineteenth century but from a global perspective on cinematography. Mimura's lighting strategy to achieve the film's depth cinematography had a dialogic relationship with Gregg Toland's deep-focus cinematography in Hollywood. The *à traverse* effect in Mimura's work was constructed in Japan as a result of his sensitivity to Hollywood technology and Japanese reality.

In *Humanity and Paper Balloons*, lighting is carefully used for a compositional purpose in order to enhance the depth of field in long shots. The set of tenement houses, where major events occur, is structured on the principle of linear perspective. Two rows of tenement houses occupy the two sides of the frame from front to back. At the far end, there is a warehouse-like building, and the streets are T-shaped there. Compared to *Citizen Kane*, however, we may not be able to call this a shot of deep focus. The far end of the screen is not in sharp focus. Yet, it is noteworthy that the T-shaped area at the back is brightly lit, which enhances the sense of depth. People who live there but do not have particular roles in the narrative walk back and forth on the street. They speak arbitrary dialogue and move similarly arbitrarily. The arbitrary movements of these people constantly block the brightest area of the screen at the far end and create visual zigzagging between the front and the back (*à traverse*).

There are major characters of this film whose fates come to tragic ends in the finale, but their movements are not captured in a particularly privileged manner. Like other people of the tenement houses, they walk on a daily basis from the frontal dark area to the brighter area at the back. As a result, their backs are often displayed in silhouette. Unno (Kawarazaki Chojuro), a masterless samurai living in one room of the tenement house, walks down the rainy street from a frontal shadowy area to a lit area at the back every time he is dismissed by Mori, a high-ranking samurai, after asking for a job.

Even inside a room, Mimura maintains his system of "compositional lighting." In a climactic scene at home, Unno's wife slowly approaches her husband, who is drunk and sleeping beside an *andon* lamp at the back of their room. When the wife shows a small dagger in her hand, it shines momentarily at a dark corner of the room, reflecting the light from the lamp. When she blows out the *andon* lamp, a window at the back becomes the only visibly lit area of the screen. The eyes of the viewers are invited to move from the back to the front to the back again, following the swift changes of the brightest spots within the same composition of the depth. Lighting is thus used to enhance the *à traverse* effect between the back and the front in this scene. The people of the tenement houses learn about the "double suicide" of Unno and his wife the following morning. The film ends with a long shot of a vacant street with a side ditch, where one of the paper balloons made by Unno's wife to earn their living, slowly floats from the frontal shadowy area to the bright area at the back.

THE CINEMATOGRAPHY OF DEPTH AND THE POST-WORLD WAR II GEOPOLITICS

Immediately following Japan's announcement of surrender to the Allied Powers on August 15, 1945, Japan was placed under a "neo-colonial military dictatorship" of American occupation, which continued until signing of the San Francisco Peace Treaty in April 1952.[36] One of the major goals of the occupation was to reconstruct Japanese politics and economy as soon as possible in order to turn Japan as the forefront of the U.S. foreign policy in Asia during the Cold War. For that purpose, the occupation government highly regarded education of Japanese people with American-style liberalism and democracy. The occupation government turned to the effectiveness of media, especially cinema, to publicize this new ideology to Japanese society at large. As early as September 1945, the occupation government met with the executives of Japanese film studios to communicate their interest in the film industry. By October 1945, the Information Dissemination Section (later the Civil Information and Education Section, or CIE) was established and started exercising censorship of film projects, scripts, and completed films by checking translated materials submitted by film companies.[37] Under the protection as well as the severe control of the occupation government, the Japanese film industry reestablished itself rapidly. Even after the occupation ended, the political and economic influence of the U.S. over Japan continued. Such post-World War II geopolitical conditions gave the cinematography of depth, which initially developed in the 1930s, and consequently the discussions on flatness, intriguing new dimensions.

First, there was a director who was obsessed with deep-focus cinematography and vigorously tried to adopt the Hollywood system. The name of the director was none other than Kurosawa Akira (1910–1988). It is imaginable that Kurosawa was initially drawn to deep-focus cinematography by way of Mimura Akira. Mimura was the cinematographer for *Sanshiro Sugata* (*Sugata Sanshiro*, 1943), the directorial debut film of Kurosawa. According to Hara Kazumi, who worked as an assistant cinematographer for seven Kurosawa films after *Seven Samurai* (*Shichinin no samurai*, 1956):

> He [Kurosawa] started thinking that everything on the screen must be seen clearly simply because he was a painter as you know. Paintings are in deep focus. His idea stemmed from it. Pictures are drawn from the edge to the edge of

> a canvas. Deep focus was the same idea. Or, it might be better to say that his idea was based on the human eyes. The eyes capture things in deep focus. Kurosawa thought that reality would be necessary in photographic images. That idea came to him because he had experiences in painting.[38]

The production technique of deep focus that Kurosawa adopted was very similar to that of Orson Welles and Gregg Toland in *Citizen Kane*. While the standard lens and f-stop at Toho studio, where Kurosawa was contracted, were 50mm and f:4 in the mid-1950s, Kurosawa preferred to use telephoto lenses of 100mm, 135mm, 250mm, or even 500mm.[39] The standard f-stops for *Seven Samurai* was f:5.6 and f:6.3, and Kurosawa requested f:8 and f:11 every once in a while.[40] The films that they used were Fuji panchromatic negative, whose ASA was 64, which was not very sensitive compared to the Hollywood standard.[41] If that was the case, his films needed much more luminosity than standard Toho films. A huge number of arc lamps and a huge amount of electricity were necessitated. The focal length was extremely limited, too. As Hollywood studios did not like Welles, Toho did not like Kurosawa for financial reasons. After the end of World War II, Japanese economy was rapidly reconstructed under the protection of the United States in the Cold War. A 1956 economic white paper by the Economic Planning Agency of the Japanese government declared, "The postwar years are no longer." The film industry was particularly placed under a serious control of the U.S. occupational government from 1945–1952. Kurosawa's extravagant cinematic experiments were possible because of such an economic boom. Moreover, after suffering from multiple labor disputes during the occupational period, Toho restructured itself as a company and was about to regain lost ground. The name of Kurosawa Akira was strong enough to place a bet on, especially when his film *Rashomon* (1950) just received the Golden Lion award at the Venice International Film Festival in 1951. (Once the film industry started to decline in the 1960s, Kurosawa was forced to separate from Toho and became independent.)

If filmmakers growingly rely on long lenses, as film scholars Kristin Thompson and David Bordwell suggest, they tend to create "a flatter-looking, more cardboardy space than wide-angle lenses do."[42] Thompson and Bordwell continue, "A telephoto shot makes an image seem less voluminous. It seems to be made of slices of space lined up one behind another," which is "well-suited to a 'painterly' or strongly pictorial approach to cinema."[43] As a result, the images of Kurosawa's films look more flat than other Japanese films of the same period. It is ironic because that happened when Kurosawa did not resort to the Japanese sensibility of superflatness that Murakami Takashi insists but because Kurosawa vigorously adopted the Hollywood technique of deep focus. Mainly because of Kurosawa's intensive use of telephoto lenses, the distance between the front and the back is shunned.

In his standard-size films, such as *Ikiru* (1952) and *Seven Samurai*, the images are captured not in a wide-angle space but look packed in and claustrophobic. Kurosawa did not include anything surplus in his images. Telephoto lenses were perfect for that purpose because of its limited scope. In *Ikiru*, for instance, Watanabe (Shimura Takashi), who has stomach cancer and less than a year to live, suddenly finds a new purpose for living after spending some time with a cheerful hardworking woman. In that scene, there is a shot that juxtaposes Watanabe at front rushing down a staircase of a restaurant with a new hope and schoolgirls at back singing "Happy Birthday" at a party. Supposedly, in such a composition of depth of a Toho film, there is the *à traverse* effect and the visual interaction between the front and the back. However, because everything visible is in focus and packed in a tight standard-size space, the schoolgirls appear to be singing the song right on Watanabe's shoulder and directly to his ears, as if they were small angels sent from heaven.

Another development on the cinematography of depth was a result of a self-conscious reference to the flatness of a Japanese aesthetics form and a strategic adaptation of it to a cinematic form. After the rather accidental success of *Rashomon* at the Venice International Film Festival, international distribution of films became a prevalent aspiration for the Japanese film industry. A major strategy taken by Japanese film companies to appeal to international audiences was to emphasize such cultural motifs as *noh* and Kabuki drama, Zen Buddhism, samurai, and geisha that were self-consciously marked as traditionally Japanese. Exotic Japaneseness was sold as a commodity to foreign audiences and publicized as a symbol of the national identity of Japan to be approved internationally. Daiei (Dai Nippon Eiga Seisaku Kabushiki Kaisha, established on January 27, 1942, combining Nikkatsu, Shinko Kinema, and Daito Eiga) initiated the exoticization of Japanese cinema under Nagata Masaichi, the president.

Under such post-World War II geopolitical conditions, at Daiei, Miyagawa Kazuo (1908–1999), the cinematographer of *Rashomon*, started to adapt the visible flatness of *emakimono*, or picture scroll, that was created during the eleventh to sixteenth centuries in Japan to his cinematography. In *emakimono*, which Murakami Takashi considers to be one of the examples of superflat, an illustrated narrative is told in a long and horizontal space.

As I have discussed elsewhere, Miyagawa was a strategic, self-conscious traditionalist.[44] Miyagawa incessantly confirmed after World War II that his work was the embodiment of Japanese beauty. In 1985, for instance, Miyagawa wrote,

> "I was born and raised in Kyoto. As you know, the objects and the landscapes of Kyoto are filled with quiet and dark colors. My eyes naturally remember them and have formed my sense of colors. Also, I learned *sumi-e* [traditional monochrome ink paintings], from my teacher Harada Korei. . . . Mr. Harada was a strange teacher and never permitted us to use colors. He kept saying, 'Use exclusively *sumi* [Japanese black ink].' *Sumi-e* is a world of black and white, and infinite levels of gray. He meant that we had to make viewers feel colors by the level of thickness [of gray]. My cinematographic tone was born thanks to my training in *sumi-e*."[45]

But he also admitted that at the base of his cinematographic style was a strong influence of Hollywood cinema. Miyagawa stated, "We cannot teach or be taught camerawork. We have to think of a camera as a part of our bodies and move it based on our own sense. There is no other way. We have to master it naturally. I watched many foreign films, American films in particular, and learned this. I learned especially from Hollywood musicals that I watched in 1939–1940."[46] His early work, such as *Singing Lovebirds* (*Oshidori utagassen*, Makino Masahiro, 1939), a musical, displays Miyagawa's strong inclination toward Hollywood-style three-point lighting instead of *sumi-e*.

Ugetsu (*Ugetsu monogatari*, Mizoguchi Kenji, 1953), which received the Silver Lion award at the Venice International Film Festival in 1953, was a good example of Miyagawa's incorporation of the "traditionally Japanese" aesthetics. Miyagawa called *Ugetsu* a film with "infinite gray as the basic tone," which could give "a sense of calmness to human eyes."[47] "In movie theaters," Miyagawa added, "gray melts into blackness that surrounds the screen. When we watch films in theaters, gray certainly appears to be spreading out of the frame into the space outside. This is what cinematographers have to think of most."[48] However, such emphasis on gradation was so nicely in tune with his thought on *sumi-e* that it seemed to be Miyagawa's official attitude to publicize his cinematography as the Japanese aesthetics for the purpose of exoticization. It was because when Miyagawa talked about the "infinite gray" of the film in an interview, he contradicted himself by saying, "Technically, I used arc light. The lighting source of arc light is a dot so that it creates a sharp shadow. . . . If we use an arc lamp for frontal light, it can make soft diffusing light."[49] Here, Miyagawa mentioned how he achieved certain soft tones, but actually he was referring to the resulting sharp shadows by use of the same lighting equipment.

In a similar manner, flatness in this film was not an embodiment of the Japanese sensibility of superflat but a result of a strategic choice. The film *Ugetsu* extensively uses long takes and long shots. A typical example is a shot/scene that depicts a tragic murder of Miyagi (Tanaka Kinuyo) by soldiers. This is a high angle extreme long shot. Miyagi carrying her child Genichi walks along a mountain road from the right. She is stopped by two solders stepping out of a hut. The soldiers steal Miyagi's food. Miyagi tries to defend herself but is stabbed with a spear of the third soldier. The crane camera follows Miyagi's desperate steps to the left. The long duration and the slow movement of the camera from right to left imitate the narrative development of *emakimono*. Moreover, when the shot becomes momentarily static at the end of the shot/scene, the composition of depth in deep focus, supposedly thanks to the use of arc lamps on this location shooting, contrasts three layers in a brutal manner: dying Miyagi in the middle plane and the soldiers in the background devouring rice cakes that they stole from her. In addition, a shrub is placed at front, through whose silhouette we see Miyagi and the soldiers. In Miyagawa's original idea, the shrub was a bigger tree standing at the edge of the Lake Biwa.[50] In this sense, the *à traverse* effect of woodblock prints is also replicated here.

CONCLUSION

On March 29, 2012, the *Asahi* newspaper digital edition reported that in principle Japanese major film companies —Toho, Shochiku, and Toei—would no longer produce any 3-D films.[51] According to the report, there were two reasons for their decision. First, they could not find audience that would watch their products even when many viewers liked to watch Hollywood-made 3-D films. Japanese audience did not want Japanese films in 3-D. The second reason was financial. One Toho producer stated that it would take a 1.5 times longer production period and a couple of hundreds million yen more production cost to make a film with 3-D cameras. When an average production cost of a Japanese film was nearly ten times smaller than a Hollywood film, it would be too risky to choose the path.

The first reason seems to fit well with Murakami Takashi's culturally essentialist theory of superflat in Japanese sensibility. In her attempt to define 3-D cinema, Miriam Ross states:

> The most dynamic and readily discernible qualities unique to stereoscopic moving images are the use of negative parallax (in the auditorium) and positive parallax (receding depth) space. In the former, there is the sense that images are placed somewhere between the screen and the place where the viewer is located. Use of negative parallax space provokes the greatest interest in popular discussion of 3D cinema and is mythologised through the many accounts of viewers seeing objects (spears, arrows, cannon balls) fly directly at them. Use of positive parallax space, which displays objects existing in deep space, behind the plane where the screen rests, is often less easy to discern as it offers a heightened presentation of the three-dimensional depth effects apparent in flat cinema. Nonetheless, it has the ability to draw attention to depth construction in a way that is not available elsewhere.[52]

Here, Ross seems to presuppose the innate linear perspective in the images of 3-D cinema. If that is the case and if Japanese audience does not want to see Japanese films in 3-D, it may be possible to argue that there is Japanese sensibility toward visual flatness.

The second reason sounds too hasty. The history shows that the cinematographers and the critics of the 1930s knew their material and financial limitations compared to Hollywood but attempted to formulate their own lighting schemes and compositions of depth. Even experiments in 3-D cinema were conducted at Toho, where Mimura Akira worked, during wartime under the supervision of Japanese Navy. The so-called To-Vision was realized almost the same time when United Artists' *Bwana Devil* (Arch Oboler, 1952) initiated the first 3-D boom in the U.S. In the 1980s in the midst of the bubble economy, Japanese technology companies, including Sony and Fujitsu, led the world in the production of 3-D films for World Expositions.[53]

Responding to the decision by the major Japanese film companies, film journalist Otaka Takao insists, "Japanese cinema should have its own way of 3-D expression even if it may not be comparable to Hollywood. Technical challenges should not be given up now."[54] As I have done throughout this essay, I want to focus more on the multiple layers and the physiological interactions among them in apparently flat-looking screen images of Japanese cinema and anime. I also want to draw attention to the conflicts and negotiations between depth and flatness in the historical development of cinematography in Japan. Both the cinematographers and the audience in Japan have not simply preferred 2-D films to 3-D as the *Asahi* newspaper report explains. They have had more complicated attitudes toward the two-dimensional material of movie screen and the three-dimensional visuality of cinema than the theory of superflat.

In the same year when the *Asahi* newspaper reported the major Japanese film company's decision, an independent 3-D film was produced and released in Japan. *Flashback Memories 3D* (2012), directed by Matsue Tetsuaki and photographed by Watanabe Tomonori, is a music documentary film about Japanese didgeridoo (Australian aboriginal instrument) player Goma, whose memory functions were severely damaged in a 2009 road accident. Goma struggles to recall his memories with his wife and child as well as his past performances. He also has difficulty maintaining new memories.

Flashback Memories 3D does not rely on the negative and positive parallaxes that Ross defines as the essence of 3-D cinema, maybe except the extreme length of a didgeridoo, which creates the "in-your-face" effect—but with no relation to the film's narrative. Instead, this film uses 3-D technology for multiple layers and their interactions. In this sense, this film can also be called *à traverse* cinema. The concert in which Goma and his band play is photographed in 3-D while animation that recreates Goma's "flashbacks" images in his brain and home video that recorded his past are inserted as well as projected onto a backdrop behind the band in 2-D. For Matsue, 3-D cinema is not an embodiment of photographic realism but "like a kaleidoscope, which the viewers look or even go into a small object and fill all their visions with the thing."[55] With the technology, Matsue claims that he wanted to enter Goma's brain and "makes this film's viewpoint itself closer to Goma's."[56] Film scholar Thomas Elsaesser insists that stereoscopy, or 3-D cinema, offers "a different kind of mental image."[57] A film like *Flashback Memories 3D*, which tried to capture Goma's "mental image," presents a unique type of visuality with the 3-D cinematography that goes beyond photographic realism or the simple dichotomy between depth and flatness.

NOTES

1. Murakami Takashi, *Super Flat* (Tokyo: Madra Publishing, 2000), 5.
2. Thomas Lamarre, *The Anime Machine: A Media Theory of Animation* (Minneapolis: University of Minnesota Press, 2009), 112.
3. Ibid.
4. Ibid., 113.
5. The multi-plane camera is a special camera used in the traditional animation process that moves a number of pieces of artwork past the camera at various speeds and at various distances from one another. This creates a three-dimensional effect. The first multi-plane camera was invented by Lotte Reiniger for her animated feature *The Adventures of Prince Achmed* (1926). In Japan, Seo Mitsuyo was the first animator to work with the multi-plane camera. In his film *Ari-chan* (Ants, 1941), commissioned by the Ministry of Education, Seo could afford to experiment with such costly techniques.
6. Thomas Looser, "Superflat and the Layers of Image and History in 1990s Japan," *Mechadmia* 1 (2006): 97.
7. Mabuchi Akiko, *Japonisumu: Genso no Nippon* (Japonisme: Japan in fantasy) (Tokyo: Brücke, 1997), 91.
8. Ibid., 100–101.
9. Looser, "Superflat and the Layers of Image and History in 1990s Japan," 101.
10. Lamarre, *The Anime Machine*, 119.
11. Aikawa Kusuhiko, "Kantokusha Yamanaka Sadao shi" (Director Yamanaka Sadao), *Kinema Junpo* 11 May 1932, reprinted in *Kantoku Yamanaka Sadao* (Director Yamanaka Sadao), ed. Chiba Nobuo (Tokyo: Jistugyo no Nihon sha, 1999), 33.
12. Ibid., 34.
13. Sugiyama Heiichi, *Eiga hyoron shu* (Collection of film criticism) (Tokyo: Daiichi geibun sha, 1941), 123, 125.
14. Sugiyama Heiichi, *Eiga no buntai: Tekunikku no densho* (The styles of cinema: Handing down techniques) (Tokyo: Koro sha, 2003), 61.

15. Kudo Miyoko, *Hariuddo kara Hiroshima e: Eiga kameraman Hari Mimura no jinsei* (From Hollywood to Hiroshima: The life of film cameraman Harry Mimura) (Tokyo: Shobun sha, 1985), 88.

16. Mimura Akira, "Gureggu tornado: Besuto iyasu no satsuei ni tsuite" (Gregg Toland: On the cinematography of *Best Years of Our Lives*), *Eiga Geijutsu* 3.7 (August 1948): 13.

17. Patrick Keating, *Hollywood Lighting from the Silent Era to Film Noir* (New York: Columbia University Press, 2010), 237.

18. Miki Shigeru, "Nihon eiga no kuraki bamen no satsuei: Rou ki ton ni tsukite ichi kosatsu" (Photographing dark scenes of Japanese cinema: A thought on the low-key tone), *Eiga Hyoron* 15.3 (March 1934): 48.

19. Soyama Naomori, "Warera no tachiba to Angusuto: Awasete Shimazaki Kiyohiko shi ni kotau" (Our position and Angst's: And response to Mr. Shimazaki Kiyohiko), *Kinema Junpo*, April 1, 1937, 113.

20. Daisuke Miyao, *The Aesthetics of Shadow: Lighting and Japanese Cinema* (Durham, NC: Duke University Press, 2013), 204–230.

21. Midorikawa Michio, "Kameraman no seikatsu to kyoyo" (Cameraman's life and culture) in *Eiga satsueigaku dokuhon: Jo kan* (Cinematography reader: Volume 1), ed. Tane Shigeru (Tokyo: Dainihon eiga kykai, 1940), 65.

22. Gregg Toland, "Realism for 'Citizen Kane,'" *American Cinematographer* 22.2 (February 1941): 54.

23. Ibid.

24. Ibid.

25. Ibid.

26. Ibid., 55.

27. Ibid.

28. Ibid.

29. Patrick Ogle, "Deep-Focus Cinematography: A Technological/Aesthetic History," *Filmmakers Newsletter* 4 (May 1971): 25–26.

30. Victor Milner, "Super xx for 'Production' Camerawork," *American Cinematographer*, November 1941, 269, quoted in Ogle, "Deep-Focus Cinematography," 25.

31. Ogle, "Deep-Focus Cinematography," 25–26.

32. Mimura Akira, "Atarashiki camera waku" (A new camera work), *Eiga Junpo* 22 (11 August 1941): 27.

33. Shimazaki Kiyohiko, "Nihon eiga no satsuei gijutsu (zoku)" (Cinematographic techniques in Japanese cinema (2)), *Kinema Junpo* September 21, 1938: 8–9.

34. Shimazaki Kiyohiko, "Eiga no gijutsu to seishin: Shiron" (Cinema's technologies and seishin: A hypothesis), *Eiga Hyoron* 2.9 (September 1942): 30.

35. Shimazaki, "Nihon eiga no satsuei gijutsu (zoku)," 8–9.

36. John Dower, *Embracing Defeat: Japan in the Wake of World War II* (New York: W. W. Norton, 1999), 81.

37. Kyoko Hirano, *Mr. Smith Goes to Tokyo: Japanese Cinema under the American Occupation, 1945–1952* (Washington: Smithsonian, 1992), 4–6.

38. Shiozawa Midori, *KUROSAWA Satsuei genba + ongaku hen: Kurosawa Akira to Kurosawa gumi, sono eiga teki kioku, eiga sozo no kiroku* (Kurosawa the set and

the music: Kurosawa Akira and his crew, the record of their cinematic memories and creations) (Tokyo: Kawade shobo shin sha, 2005), 128.

39. Ibid., 129, 139.

40. Ibid., 262.

41. Ibid., 263.

42. Kristin Thompson and David Bordwell, "Shot-consciousness." http://www.davidbordwell.net/blog/2007/01/16/shot-consciousness. Accessed March 26, 2016.

43. Ibid.

44. Miyao, *The Aesthetics of Shadow*, 255–281.

45. Miyagawa Kazuo, *Watshi no eiga jinsei 60nen: Kyameraman ichidai* (My sixty-year life with cinema: A life of a cameraman) (Tokyo: PHP kenkyujo, 1985), 195.

46. Miyagawa, *Watshi no eiga jinsei 60nen*, 38–39.

47. Ota Yoneo, "Monokuromu no jidai" (The period of monochrome) in *Hikari to kage no eiga shi: Satsuei kantoku Miyagawa Kazuo no sekai* (Film history of light and shadow: The world of cinematographer Miyagawa Kazuo), ed. Uekusa Nobukazu (Tokyo: Kinema Junpo sha, 2000), 23; Miyagawa, *Watshi no eiga jinsei 60nen*, 82.

48. Oguri Shigeki. "Renzu no mushi" (Bug of a lens) in *Hikari to kage no eiga shi: Satsuei kantoku Miyagawa Kazuo no sekai* (Film history of light and shadow: The world of cinematographer Miyagawa Kazuo), ed. Uekusa Nobukazu (Tokyo: Kinema Junpo sha, 2000), 76.

49. Ota, "Monokuromu no jidai," 23.

50. "Miyagawa Kazuo Kikigaki" (Miyagawa Kazuo interview), unpublished document, the Museum of Kyoto, 50.

51. Nishida Kensaku, "3D, hoga kara kieru? Shukyaku ni kusen, shigatsu iko wa zero" (Will Japanese 3-D films disappear? It is difficult to find audience, and there will be no film after April), *Asahi Shimbun Digital*, 29 March 2012: n.p. http://www.asahi.com/showbiz/movie/TKY201203290188.html. Accessed March 18, 2016.

52. Miriam Ross, *3D Cinema: Optical Illusions and Tactile Experiences* (London: Palgrave Macmillan, 2015), 5.

53. Oguchi Takayuki, "Dai 3 no kakumei: Rittai 3D eiga no jidai (7) Nippon no 3D eiga" (The 3rd revolution: The period of 3D cinema (7) 3D cinema in Japan), *Eiga.com*, May 26, 2010. http://eiga.com/extra/oguchi/7. Accessed March 18, 2016.

54. Nishida, "3D, hoga kara kieru? Shukyaku ni kusen, shigatsu iko wa zero," n.p.

55. "*Flashback Memories 3D* taidan Matsue Tetsuaki (Kantoku) x Takane Junji (Purodyusa)" (Conversations between Matsue Tetsuaki (Director) and Takane Junji (Producer)), *CINRA.STORE*, January 18, 2013, http://store.cinra.net/feature/detail.php?postid=220. Accessed March 28, 2016.

56. Ibid.

57. Thomas Elsaesser, "The 'Return' of 3-D: On Some of the Logics and Genealogies of the Image in the Twenty-first Century," *Critical Inquiry* 39.2 (2013): 218.

BIBLIOGRAPHY

Chiba, Nobuo, ed. *Kantoku Yamanaka Sadao* (Director Yamanaka Sadao). Tokyo: Jistugyo no Nihon sha, 1999.

Dower, John. Embracing Defeat: Japan in the Wake of World War II. New York: W. W. Norton, 1999.

Elsaesser, Thomas. "The 'Return' of 3-D: On Some of the Logics and Genealogies of the Image in the Twenty-first Century," *Critical Inquiry* 39.2 (2013): 217–246.

Hirano, Kyoko. *Mr. Smith Goes to Tokyo: Japanese Cinema under the American Occupation, 1945–1952.* Washington: Smithsonian, 1992.

Keating, Patrick. *Hollywood Lighting from the Silent Era to Film Noir.* New York: Columbia University Press, 2010.

Kudo, Miyoko. *Hariuddo kara Hiroshima e: Eiga kameraman Hari Mimura no jinsei* (From Hollywood to Hiroshima: The life of film cameraman Harry Mimura). Tokyo: Shobun sha, 1985.

Lamarre, Thomas. *The Anime Machine: A Media Theory of Animation.* Minneapolis: University of Minnesota Press, 2009.

Looser, Thomas. "Superflat and the Layers of Image and History in 1990s Japan," *Mechadmia* 1 (2006): 92–109.

Mabuchi, Akiko. *Japonisumu: Genso no Nippon* (Japonisme: Japan in fantasy). Tokyo: Brücke, 1997.

Miyagawa, Kazuo. *Watshi no eiga jinsei 60nen: Kyameraman ichidai* (My sixty-year life with cinema: A life of a cameraman). Tokyo: PHP kenkyujo, 1985.

Miyao, Daisuke. *The Aesthetics of Shadow: Lighting and Japanese Cinema. Durham,* NC: Duke University Press, 2013.

Murakami, Takashi. *Super Flat.* Tokyo: Madra Publishing, 2000.

Nishida, Kensaku. "3D, hoga kara kieru? Shukyaku ni kusen, 4-gatsu iko wa zero" (Will Japanese 3-D films disappear? It is difficult to find audience, and there will be no film after April), Asahi Shimbun Digital, March 29, 2012. http://www.asahi.com/showbiz/movie/TKY201203290188.html. Last accessed March 18, 2016.

Ogle, Patrick. "Deep-Focus Cinematography: A Technological/Aesthetic History," *Filmmakers Newsletter* 4 (May 1971): 19–31.

Oguchi, Takayuki. "Dai 3 no kakumei: Rittai 3D eiga no jidai (7) Nippon no 3D eiga" (The 3rd revolution: The period of 3D cinema (7) 3D cinema in Japan), *Eiga.com,* 26 May 2010. http://eiga.com/extra/oguchi/7. Last accessed March 18, 2016.

Ross, Miriam. *3D Cinema: Optical Illusions and Tactile Experiences.* London: Palgrave Macmillan, 2015.

Shiozawa, Midori. *KUROSAWA Satsuei genba + ongaku hen: Kurosawa Akira to Kurosawa gumi, sono eiga teki kioku, eiga sozo no kiroku* (Kurosawa the set and the music: Kurosawa Akira and his crew, the record of their cinematic memories and creations). Tokyo: Kawade shobo shin sha, 2005.

Sugiyama, Heiichi. *Eiga hyoron shu* (Collection of film criticism). Tokyo: Daiichi geibun sha, 1941.

———. *Eiga no buntai: Tekunikku no densho* (The styles of cinema: Handing down techniques). Tokyo: Koro sha, 2003.

Tane, Sshigeru. (ed). *Eiga satsueigaku dokuhon: Jo kan* (Cinematography reader: Volume 1). Tokyo: Dainihon eiga kykai, 1940.

Thompson, Kristin & Bordwell, David. (2007) "Shot-consciousness." http://www.davidbordwell.net/blog/2007/01/16/shot-consciousness. Last accessed March 26, 2016.

Uekusa, Nobukazu, ed. *Hikari to kage no eiga shi: Satsuei kantoku Miyagawa Kazuo no sekai* (Film history of light and shadow: The world of cinematographer Miyagawa Kazuo). Tokyo: Kinema Junpo sha, 2000.

Chapter 4

The Eyes of the World

Christopher Doyle, Anthony Dod Mantle, Roger Deakins and the Emergence of a Transnational Cinematic Language

Evan Lieberman

From its etymological roots in Greek "cinematography" means writing (or recording) with movement, and if it is indeed a type of writing then it stands to reason that there must be some operative language from which it draws its signifying material and within which it works to create meaning. However, the debate over whether there exists a language of the cinema at all has long been one of the most hotly contested issues in the field of cinema studies. Even pioneering cine-semiotician Christian Metz rejects the notion of film as a language system though it might "be considered as a language to the extent that it orders signifying elements, within ordered arrangements different from those of spoken idioms." (71) So as to establish a more tangible semiotics of cinema, Metz chose to focus on the syntagmatic axis of signification or editing and sequencing much like the Soviet theorists of the 1920s. For Metz, "the paradigmatic category in film is condemned to remain partial and fragmentary, at least as long as one tries to isolate it on the level of the image," (70) but it is exactly the meaning of the image, of light, color, composition, movement, and space, that is at the heart of cinematic signification. It is these elements of the image that are of central concern for the cinematographer and the student of cinematography, and that are fundamental to any examination of the language of cinema, a language that like all living languages is constantly evolving, developing, growing in order to most effectively communicate the ideas of its time.

In a paper delivered at the 2014 University Film and Video Association Conference I staked out a claim for the idea that the language of cinema was not solely the product of directorial vision nor of producers competing for audience in the early decades of the motion picture, but rather the result of a collaborative dialectic between directors and directors of photography (or cameramen as they were then known). Beginning with D.W. Griffith and Billy Bitzer and continuing through such important pairings as Charlie Chaplin and Rollie Totheroh, Buster Keaton and Elgin Lessley, Sergei Eisenstein and Eduard Tisse, and F.W. Murnau and Karl Freund, the period from roughly 1908 until the coming of sound saw a veritable explosion in the aesthetic, and communicative potentials of the moving image. Whether Griffith and Bitzer's revolutionary use of camera angle and varying shot scales, Keaton and Lessley's use of the camera as a machine for creating the physically impossible, Eisenstein and Tisse's powerful iconic imagery and symphonic compositions, or Murnau and Freund's expressively unchained camera, it was the collaborative relationship between director and cameraman that moved cinema from its theatrical Primitive Mode of Representation (Burch, 1990) to a mature art form. The more detailed specifics of this argument will have to await further revision before appearing in print, but the contention that the moving image language is in large part a product of the creative interaction between director and director of photography can be definitively established. This phenomenon should not be seen as being isolated to the early days of cinema but to the contrary that which has continued unabated throughout the development of the medium, remaining as true for contemporary filmmaking as it was when Griffith and Bitzer were experimenting with "object shots" and "Rembrandt lighting" between 1910 and 1913.

In one of the most important essays positing the existence of a moving image language, "The Evolution of the Language of Cinema" (written between 1950 and 1955 but not published in English until 1967), Andre Bazin points to the long-shot long take aesthetic that can be found in the work of directors Jean Renoir, William Wyler, and Orson Welles as an epochal shift in the methods of cinematic representation. Bazin writes that, "depth of field is not just a stock in trade of the cameraman like the use of a series of filters or such-and-such style of lighting, it is a capital gain in the field of direction—a dialectical step forward in the history of film language." (49) It is notable here that Bazin sees the move toward composition in depth as belonging to the "field of direction" and most explicitly not to that of the cameraman, particularly given the fact that the Welles film to which he devotes the most discussion, *Citizen Kane* (1941) and the Wyler films he so greatly admires are the work of the same director of photography, Gregg Toland, whose signature deep focus style is analyzed in greater depth in "Authors of the Image: Gregg Toland and Gabriel Figueroa" (2006), which I coauthored with Kerry Hegarty. However, even a cursory examination of the body of films shot by Toland and those directed by Welles and Wyler but not photographed by the master of deep-focus, reveals that the style so effusively praised by Bazin is actually far more indebted to the work of the cinematographer than the director.

Bazin does write that, vis-à-vis the films of Wyler and Welles, "[b]etween them, director and cameraman have converted the screen into a dramatic checkerboard, planned down to the last detail." (49) Here Bazin seems to be indicating some recognition of Toland's participation in the visual design of these films even if throughout the essay, he neglects to mention his name.

Bazin's essay was written during the years preceding the next great stride in the development of the cinematic language, which was connected to the rise of the international art cinema in the 1950s–1970s. The often radical work produced by such director/director of photography teams as Ingmar Bergman and Sven Nykvist (and Gunnar Fischer before Nykvist), Federico Fellini and Otello Martelli, Jean-Luc Godard and Raoul Coutard, Alain Resnais and Sascha Vierny, and Bernardo Bertolucci and Vittorio Storraro. signaled the move toward a modernist film language, one that was more personal, more abstract, figurative rather than literal, overtly metaphorical and challenging to the viewer but nonetheless authentically expressive of the cultural shifts occurring during this period. The tremendous importance of this period clearly demands its own in depth analysis, but it is interesting to note that at the conclusion of the art film era in the 1970s, cinematography began to internationalize in a significant way at least in the Hollywood film industry. Nykvist came to the United States where he worked with a variety of directors most notably Woody Allen on *Crimes and Misdemeanors* (1989), *Another Woman* (1988), and *Celebrity* (1998), as did Storaro whose endlessly inventive photography for director Francis Ford Coppola on *Apocalypse Now!* (1979), *One From the Heart* (1981), and *Tucker: The Man and his Dream* (1988) set new standards for Hollywood cinematography. On the whole this period was very much centered on international cinematographers coming to Hollywood driven by the pursuit of state-of-the-art technology, more leisurely production schedules, and of course larger paychecks, but the type of close director/director of photography collaboration that leads to real advancement of the cinematic language was not much in evidence.

The contemporary cinema is being increasingly described as transnational, and Elizabeth Ezra and Terry Rowden write in the introduction to their edited volume *Transnational Cinema, The Film Reader*, "[t]he global circulation of money, commodities, information, and human beings is giving rise to films whose aesthetic and narrative dynamics, and even the modes of emotional identification they elicit, reflect the impact of advanced capitalism and new media technologies as components of an increasingly interconnected world-system."(1) This change in "aesthetic and narrative dynamics" points toward another potential period of development in the language of the cinema, and given the way this process has worked in the past, asks the question, are there yet again director/director of photography teams working at the forefront of this movement? Complicating the issue is the fact that the relative instability of the film industry has contributed to the near impossibility of directors and directors of photography working together in the kinds of sustained relationships we see with Chaplin and Totheroh or Bergman and Nykvist. Nevertheless we can identify some pairings that can be seen as contributing to the ongoing transformation of the cinematic language in the current transnational industry, and interestingly enough, these teams are themselves from different nationalities.

Australian-born cinematographer Christopher Doyle is best known for his revolutionary work with Hong Kong director Wong Kar Wei and did not shoot his first film in the country in which he was born and raised until the 2002 feature *Rabbit Proof Fence,* nearly twenty years after his debut on the Taiwanese Edward Yang directed *That Day, On the Beach* (1983). Anthony Dod Mantle was born in Great Britain but attended the National Film School of Denmark before becoming the most prominent director of photography in the Dogme 95 movement, particularly notable for his collaboration with directors Thomas Vinterberg and Lars Von Trier. A decade after the beginning of his career he shot his first British project, *Vacuuming Completely Nude in Paradise* (2001) with director Danny Boyle. The first fifteen years of Roger Deakins' career were spent making films in his native England, but beginning with *Air America* in 1990 he has worked predominantly in the United States, shooting nearly all of the Coen brothers' films from *Barton Fink* (1991) through their most recent work *Hail Ceaser* (2016). All three of these directors of photography have worked with a wide array of directors and in ever shifting national and transnational contexts. What connects them is the fact that through their collaborations with directors of divergent nationality they are contributing to the development of a transnational language of cinema

THE WILD STYLE OF CHRISTOPHER DOYLE

Of all the figures in our study, Christopher Doyle may be the most fundamentally transnational. As virtually every biographical source details, he left his native Australia to dig wells in India, serve in the Norwegian merchant marine, herd cows on an Israeli kibbutz, attend the University of Hong Kong, and practice traditional Chinese medicine in Thailand before moving to Taipei, Taiwan where he shot his first film. Doyle was an international man before he ever picked up a camera. The stifling structure of the Taiwanese industry, saw him first emigrate to France to shoot *Noir et Blanc* (1986) before returning to Hong Kong where his career really began. His first film with director Wong Kar Wei, *Days of Being Wild* (1991), was both unpopular domestically and little noticed in the West at the time of its release but viewed from the contemporary perspective it certainly explodes off the screen with a vibrant energy and fertile visual imagination. Even from within the hyperkinetic and intensely stylized Hong Kong Wild Style that was already in full bloom by the time Wong and Doyle made *Days of Being Wild* their film looked like nothing else that had been produced by that prolific movement. It established a paradigmatic approach to the image that would remain in effect throughout their collaborative relationship even as their style grew and became somehow simultaneously more experimental and more subtle.

The signature trademarks are all here in this initial collaboration, from the searching camera that utilizes unusual multi-axis movements, to finding the most engaging and meaningful compositions, to the painterly geometry that is the product of unusual, often extreme, angles, both horizontal and vertical, and an integration of the set into the composition that emphasizes shape, line, and mass over object identification. The often uncomfortably close "dirty" close-ups that tend to feature elements of the set or other characters as soft-focus framing and masking devices (which is what makes them "dirty" as opposed to a "clean" close-up that features only the subject in the foreground of the frame) are already in evidence as are the aesthetic manipulations of color, from washed out to practically blooming, and frame rate, with frequent slow motion used for dramatic emphasis. Tony Rayns in one of the few English-language reviews to appear on the film's initial release wrote in *Time Out London* that "the swooningly beautiful camera and design work takes its hallucinatory tone from the protagonist's own uncertainties," recognizing one of the most important characteristics of the Doyle/Wong collaboration, that every element of what might appear as stylistic excess is connected to the characters and the drama in which they find themselves.

The tightly framed extreme close-ups that weave around Yuddy (Leslie Cheung) and Li-Zhen (Maggie Cheung) in the opening seduction scene pull the viewer into an intimate relationship with the characters from the very beginning and implicates the spectator in the almost claustrophobic quality of their relationship. The continuous movement of the camera, so unusual to see in close-ups prior to this film, hints at both the instability to come and Yuddy's relentless pursuit, which relies on a variety of tactics to persuade Li-Zhen to relent. The camera stops only when he has succeeded in penetrating her emotional armor. When she steps out into the hallway to watch him leave, an odd rippling blue light shimmers against the back wall in the distance, as though her hopes of a tranquil relationship are but the glimmers of a distant dream. The hallway will become another important trope in the Wong/Doyle style. Subsequent encounters between the two utilize a similar moving close-up photographic approach, only each scene changes slightly to reflect the changing nature of the relationship. Later in the film, the fairly extreme high angle shot of Tide (Andy Lau) in his police uniform standing in the rain framed by the oblique angle of a building and the water-slicked streets on one side and strong verticals of another building's wall and a telephone booth on the other side, speaks to his lonely sadness and his isolation even as it traps him within the strange geometry of the shot. Once he falls in love with Li-Zhen who is emotionally devastated due to being cast aside by Yuddy the film returns to this shot only now much closer to the subject in order to convey his increasing anguish brought on by his unrequited love. He is still lonely, isolated, and trapped but now he's hurt as well. This type of recurrent motif can be found with great regularity across the work of Wong and Doyle.

After the personal intimacy of *Days of Being Wild*, Doyle and Wong embarked on quite a different project, *Ashes of Time* (1994), a time travelling experimental martial arts epic that allowed them to further explore the boundaries of cinematic representation. In addition to the lush saturated colors of the desert, and a way of filming action sequences so that the characters and camera move past, across, around, toward, and away from one another in what amounts to a cinematic martial arts dance, the duo reveal another color within their representational palette, restraint. In one of the film's pivotal scenes Maggie Cheung playing the sister-in-law of Ouyang Feng (Tony Leung) speaks painfully about travelling through time in a static medium close-up that runs for well over two minutes. The lighting is simple but exquisite and the layers of dark geometric construction with a single slash of light on the furthest back wall provide a sense of depth and visual interest to the shot. During this most emotional moment in the film Wong and Doyle are careful not to intrude aesthetically but rather to support Cheung's powerful performance with a shot that allows her the dramatic space for expression.

Given the commonness of the sword fight in the Hong Kong Wild Style, Doyle is compelled to find a fresh tactic, using strongly composed canted angle close-ups of the combatants and a new technique for manipulating the action, step printing, in which the film is shot at a lower frame rate, for example, 8 fps but then each frame is printed three times creating an illusion of sped up action that is actually moving in "normal" time. It gives the action a unique sense of pace. In a 2008 interview with Tom von Logue Newth Wong explains, "Jackie Cheung who's the guy who came from the countryside, he's fresh, so we shot it with a different method: we used the step-printing to make it like it's very fast, you know; we want to tell you he's speedy, he's very fast." Wong and Doyle would return to the step printing process many more times over their collaborative careers, most notably in the film they made during the break between shooting and editing *Ashes of Time*, *Chungking Express* (1994), the film that truly solidified the style of Doyle and Wong and positioned them among the most admired and celebrated filmmakers of the contemporary cinema.

Chungking Express begins with a bravura sequence of a cop, He Qiwu (Takeshi Kaneshiro), chasing a criminal through the labyrinthine Chungking

Mansion. The stuttered energy of the step printing process, combined with the maze of corridors, the mixed color temperature lighting, and the handheld follow shot give the sequence a unique intensity unlike anything previously seen on screen. The point however is not simply to generate excitement but rather to contrast the violence of the high-speed chase with the stillness and somber tone of the images that follow: the static close-up of the clock, clouds passing by the expressionistically skewed buildings against an unnaturally blue-gray sky, and perhaps most significant, another high angle shot of a lonely cop standing underneath a streetlight (so similar to the shot in *Days of Being Wild*). The real story is not the life-and-death struggle of police work that served as the basis for so much of post-John Woo Wild Style cinema, but rather an obsession with time in the years leading up to Hong Kong's return to Chinese control, the idea that nature is dark but relentless, and that ultimately what matters to an individual has nothing to do with politics or profession but rather with emotion.

The film is a riot of color and movement. The extraordinary handheld camera that follows Faye (Faye Wong) on her illicit visits to the apartment of Cop 663 (Tony Leung) as she cleans his place, adds goldfish to his tank, and revels on his bed have the quality of perfectly improvised jazz. The camera in these sequences remains in near constant motion but never in ways that are predictable. It bobs and ducks, follows in sweeping motions and short bursts, moving with the character, past the character, over and around the rooms, and in every way conveying the giddy excitement and edge of danger that Faye is experiencing while occupying this forbidden space. There is very little in the way of what might be called naturalistic lighting (though of course it is entirely artificial) in the film and scenes are colored like a moving image painting. When He is drinking his sorrows away at The Bottoms Up Club, there is an unusual interplay of green and orange on his face. When The Woman in Blonde Wig (Brigitte Lin) is introduced, she is in a warm orange/amber. At the point that He goes to sit next to her he still has a green side to his face, while she somehow avoids the green. Once he comes to terms with her romantic and sexual disinterest he loses the green as well though it should be noted that it still appears on a curtain the background. Symbolism of color has been a signifying element of color cinematography since it began, and Jack Cardiff and Michael Powell (and Emeric Pressburger) mastered its emotive and thematic powers as early as *Black Narcissis* in 1947. Wong and Doyle's use of color is different, bolder, more assertive, more abstract and conceptual, with the image becoming a canvas on which to paint either monochrome like the cool blue of the inside of the Circle K where He buys the outdated pineapple or multihued like the exterior of the building, where our heartbroken cop offers an outdated can of pineapple to a vagrant, who turns him down.

Fallen Angel repeats many of the motifs from *Chungking Express*. Killer's Agent (Michelle Reis) is so in love with the hitman for whom she works that she cleans his apartment, though now there are strongly mixed color temperatures and the movements are generally shorter, punchier, more constrained. She is more methodical as contrasted with the adrenaline fueled thrill of Faye Wong's breaking and cleaning in the previous film. When Wong, the hitman, is leaving his apartment it is step printed but here it's not about the intensity of kinetic energy but rather an inner tension. From their first film together, Wong and Doyle have been visually fixated on corridors, a propensity that finds it greatest realization in the early sequences of *In the Mood For Love* (2000). The apartment corridor in this film gives Doyle and Wong so many visual opportunities—there are the converging lines as the eye moves down the hallway that allows for the kind of geometry of composition they prefer, provides multiple planes of action, and allows for zone oriented, often mixed color lighting. Here the hallway is a common space, a place of passing and of meeting, a place where the unexpected might happen. *2046* (2004) is the last film on which Doyle and Wong have collaborated as of this writing, and though two other directors of photography are credited on the film (Kwan Pun-leung and Lai Yiu-fai), in terms of its sumptuous color palette, ceaselessly inventive camera movement and composition and intricacy of visual design it can be seen as something of a culmination of the partnership.

Even with all of the cinematic techniques innovated or employed by Wong and Doyle perhaps the most significant contribution made by the two to the emerging contemporary film language is the notion that every shot in the films made by the pair is absolutely unique. There is virtually no such thing as a standard close-up or medium shot. Every image is specifically and often spontaneously composed for a layered effect of affect and meaning. The visually stunning, oddly rolling flyover shots of the Argentinian Iguazo waterfalls in *Happy Together* have a powerful, stirring emotional charge even as they evoke the tumultuousness and violence of the relationship between Ho (Leslie Cheung) and Lai (Tony Leung). The audience has certainly seen such aerial nature shots before but never ones that look like these. The frequent lingering "dirty" close-ups of Tony Leung in *In the Mood For Love* and in its sequel of sorts *2046* are completely different with the latter being more somber, pensive, foreboding as a result of small changes in composition, a different color palette and degree of saturation, and of more restrained camera movement. Even within the respective films each shot is distinct.

It would not be incorrect, or overstating, to say that Wong and Doyle have developed a distinctive cinematic vocabulary, their own language in some sense through which they can tell their stories of transnational alienation, longing for love, and a desire for a connection to places that may exist only in the mind. The lexicon they have developed, the dynamic "dirty" close-up, the searching camera, the use of mixed color temperatures, image manipulation through slow motion, step-printing, and ramping the camera (changing the frame rate in the midst of the shot), the dramatic use of corridors, the unique quality of every image, all of these have become part of the broader language of cinema. All of these have changed the way that filmmakers write in motion. In an interview printed in *FilmCraft: Cinematography* Doyle has said, "The function of the cinematographer is to be the bridge, the conduit between the audience and what's in front of the camera. To me there are only three people in cinema: the person in front of the camera, the audience member, and the person who is the real passage between them—the cinematographer."[1]

THE DV AESTHETIC OF ANTHONY DOD MANTLE

Anthony Dod Mantle's cultural background might not be as multinational as Christopher Doyle's, but like Doyle he left his home (England in this case) travelled (primarily through India), and wound up living and working in a national context different from that in which he was born and raised (Denmark). Also while Doyle came to prominence primarily through his work with Wong Kar Wei, it was Dod Mantle's connection to a movement that led him to develop his signature style of cinematography and so any analysis of that style has to include his continuing work with two directors, Thomas Vinterberg and Lars von Trier. Vinterberg and von Trier are the co-founders of the Dogme 95 movement, which began as something of a joke before emerging as perhaps the most important film movement since the Hong Kong Wild Style in the 1980s. To a large degree the cinematography of the Dogme films were dictated by the Manifesto or "Vows of Chastity" that Vinterberg and von Trier wrote (allegedly in forty-five minutes) and had printed into pamphlets with which they bombarded the attendees of *Le Cinéma Vers son Deuxième Siècle* (cinema toward its second century) conference in Paris in March of 1995. Rules 3, 4, and 5 from the vows are particularly significant in terms of cinematography. First, "the camera must be hand-held. Any movement or immobility attainable in the hand is permitted." Also in what really amounts to two separate rules, "The film must be in colour. Special lighting is not acceptable. (If there is too little light for exposure the scene must be cut or a single lamp be attached to the camera,)" And finally, "Optical work and filters are forbidden" (in Utterson, p. 88). Clearly the imposition of such limitations dictates in many ways what the director of photography may bring to the film, or in this case may not bring to the film, including such fundamentals as tripods, dollies, lighting instruments, gels, and filters. By stripping away these basic tools of filmmaking though, the Dogme vows enabled the development of a new approach to cinematography, one rooted deeply in character, drama, and the real.

During the ten years of the movement's official existence all certified Dogme films were numbered, and Dogme #1 released in 1998 was *Festen (The Celebration)* directed by Vinterberg and photographed by Dod Mantle. In many ways *Festen* represents the blueprint for Dogme style cinematography and sets the stage for a decisive shift in the language of the camera through the lens of Anthony Dod Mantle. When Dod Mantle shot *Festen* digital video (DV) was in its infancy having only been commercially introduced in 1995. Rather than try to mask the visual limitations of the format due in part to its lossy compression, limited dynamic range, and non-interchangeable zoom lens, he incorporated these features into the style of the film. The intensely blown-out windows in the interiors, particularly during Christian's speech go beyond overexposure, often blooming beyond the outlines of the window frames. Though inevitable when the "rules" one is working under forbid putting neutral density gel on the windows or artificially lighting the interior which under traditional shooting conditions would have created a balanced exposure, the effect works to increase the tension of the scene and represent the emotional intensity of the characters even as it adds to the visual claustrophobia that keeps the viewer and the guests trapped in the uncomfortable space of the dining room.

Because of the demands of the Dogme "vows" in the early films of the movement Dod Mantle had to find the position of the camera that would create a dimensional and thematically appropriate lighting; this despite the fact that in *Festen* he "basically told the actors to go wherever they wanted, which was very confusing for them and rather amusing to watch. . . . I would deliberately place myself in near impossible situations to see what kind of energy that created" (from *Exposure-the Magazine*). This composition on the fly so to speak is very much akin to the approach used by Doyle and Wong, and like the work of the previously discussed pair, the result is often extraordinarily evocative and aesthetically beautiful imagery. When Christian (Ulrich Thomsen) delivers the speech that is so crucial to the narrative he stands at one end of a long table of guests, not merely the focal point of the shot's vectors but also with a strong key to fill ratio that reveals his emotional division about outing his much admired father as a pedophile. By positioning the father, Helge (Henning Moritzen), at the far end of the table directly opposite his son, Dod Mantle is not only able to place him on the "hot seat" as a result of light coming in from a large window but also to create a strong composition that has Christian spatially confronting his father.

In *Festen*, as in Dod Mantle's subsequent work, the camera is in nearly constant motion, but it is different from Christopher Doyle's which seems to be alternately seeking the most perfect composition for the moment while conveying the tone and mood of the scene. Rather, Dod Mantle's camera often appears to be searching for his subject, for what is most important to shot, for the most telling detail, which occasionally he does not find, leaving something like a visual ellipses at the end of the shot. Dod Mantle's images embody the intensity of the on-screen conflict not in the same way as Doyle's dancer but more like a boxer looking to land a dramatic blow. This radical instability can be seen in the scene in which sister Helene's (Paprika Steen) black boyfriend Gbatokai (Gbtokai Dakinah) arrives and is confronted by her racist brother, Michael (Thomas Bo Larsen). The camera movements become increasingly violent becoming particularly agitated when Helen and Gbatokai kiss. When Michael tries to physically attack Gbatokai the camera essentially participates in the fight, whipping, jittering, lunging and retreating in firm contrast to the end of the scene when Helene and Gbatokai go inside leaving Michael in the driveway as a relatively static image observes Michael as his anger dissipates. In sum, *Festen* is something of a textbook of Dod Mantle's Dogme-driven innovations in the camera language but as the cinematographer himself claims, "I don't like being called 'Mr. Dogme.' The soul, not the style, is what matters."[2]

Dod Mantle's subsequent collaborations with Vinterberg abandoned the Dogme "vows" but have not produced the striking results of their initial work together. The five years in production and almost universally maligned *It's All About Love* (2003) has some exceptional use of light and color particularly in the ice skating sequences but is otherwise fairly unexceptional. *Dear Wendy* (2003) and *A Man Comes Home* (2006) similarly do not extend the innovative style seen in *Festen* though *A Man Comes Home* certainly represents something of a return to form for the pair in terms of the film's cinematography. In fact it is Dod Mantle's work with the other founder of Dogme 95, Lars von Trier, that represents his continued development of the cinematic language. The pair first worked together on an audacious experimental film for Danish television, *D-dag* (2000) for which seven Danish networks ran four films simultaneously, each representing a different point of view on the same action (a group of people using the New Year's festivities to mask a bank robbery), with each viewer using his or her own remote control as an editing device. Von Trier and Dod Mantle's episode, *Lise*, has the look of Dogme film, but by the time the pair collaborated on a feature film, *Dogville* (2002) they had clearly moved beyond the limitations of the "vows."

Both *Dogville* and its more effects intensive follow up *Manderlay* (2004) are radical departures not only for von Trier and Dod Mantle but for filmmaking in general. Shot on sound stages with only the barest suggestion of actual spaces and objects both films feature theatrical lighting, overtly artificial setting and staging, and the use of overhead camera arrays (far more extensive in *Manderlay*) the images from which can be digitally "stitched" together to create the film's "shots." The lighting effects in the far more minimal *Dogville* are often impressive, particularly in the evocation of the movements of clouds across the set, and the for his work on the film Dod Mantle won a Best Cinematography award at the European Film Awards in 2003. While it is true that in these films von Trier and Dod Mantle create a visual style unlike virtually anything ever seen before in cinema, it is hard at this point to see their work here as an advancement of the language of cinema. To the contrary, the films bear a stronger resemblance to the pre-Griffith/Bitzer work of Melies and Edwin S. Porter, a return to cinema's roots in theater and stage-bound spectacle only now fortified by motion control cameras, green screen compositing, and digital technologies. The real brilliance of von Trier and Dod Mantle's collaborative vision can be seen far more clearly in their next film together *Antichrist* (2008).

From the beautifully stylized yet emotionally harrowing opening black-and-white sequence, with its simple but sculptural lighting, exquisite composition, and ever shifting and searching focus, it is evident that the film is something of a pinnacle of achievement at this point for the von Trier/Dod Mantle pairing. Freed from both the confines of their previous films' blatantly artificial sets and the restricting rules of Dogme 95, it is a work in which the stunning cinematography is matched only by the relentless brutality of the narrative thus creating a nearly unbearable tension in every frame. This technique can be seen in the shots showing He (Willem Dafoe) and She (Charlotte Gainsbourg) making love while their infant son, Nick falls out of a window to his death. The extremely intimate close-ups convey their passion, while the quizzical expression on the young boy's face as he falls to the ground makes it appear that he is joyfully flying, though the audience is well aware of the painful truth. The contrast between passion and beauty and emotional devastation is present throughout the sequence and reveals how von Trier and Dod Mantle are writing a complex and layered imagery throughout the film.

Nowhere is this complexity more evident than in the scene in which She accuses He of planning to leave her and eventually smashes his testicles with a large piece of wood. The light is hard, harsh, minimal and unforgiving, the color palette similarly reduced to black, blue, and perfect skin tones, and the camera is in the middle of the struggle its frantic, irregular movements positioning the viewer within the struggle, not as an observer but as a horrified participant. There is a similar affect later in this section when He (capitalized because that is the only name the film gives him) has buried himself under the base of a tree and She has come looking for him to either have sex with him, kill him, or both. As He tries to unloose the grindstone that She has drilled through his leg there are several short, sharp zooms in and out and angular movements that evoke the desperation of the character as He tries to escape his cruel fate.

As difficult as *Antichrist* might be to watch at times and as polarizing as it is for audiences, it is nonetheless a film of unforgettable images and virtuoso camera work. The slow-motion crossing of the small bridge at night is done in extreme long shot with an extremely bright light source causing the water beneath the bridge to glow an otherworldly white and the background of the image to be bathed in an ethereal yet evil fog. This atmospheric diffusion serves as a significant visual motif, and when She leaves the house half naked her shirt streaked with blood after having drilled the grindstone through his leg, the air has a dark mystical weight from the heavily backlit haze. This powerful evocation of the evil surrounding them is contrasted with a flashback in which She sits in a field near the house next to Nick blowing the tendrils off a dandelion puff. The image is bathed in a golden tint and the soft-focus elements of both the foreground and background create an elegiac feeling that momentarily harkens back to a time before the tragedy, before their lives were plunged into darkness and chaos. Over and over throughout the film von Trier and Dod Mantle provide us images layered with beauty and horror, passion and pain, the sacred and the profane, bringing a complexity to the image that extends far beyond the conventions of the cinematic language.

There is no question that Dod Mantle's current reputation is built more on the films he has shot for director Danny Boyle (and more recently Ron Howard), and his academy award for the Boyle-directed *Slumdog Millionaire* has placed him in the upper echelon of international directors of photography today. However the style he has employed, most notable in the Boyle-directed films is directly drawn from his work with the Dogme founders, Vinterberg and von Trier. His understanding of light is informed both by his being denied the use of traditional instruments in his Dogme work and the radical experiments he undertook with von Trier. His use of the camera has almost always revealed the traces of a DV heritage (though his early work is on 16mm and 35mm as well as Betacam video), with its freedom of movement, searching focus, and live, indeed in many ways living, zooms. There is an unpredictability about the camera that might be seen as demanding greater attention to the image so that the viewer may try to follow it. Of all the three cinematographers discussed Dod Mantle has been most explicit about the idea of contributing to the cinematic language, as he has said, "Technique that does not become art is of no interest. People fear the new and unpredictable. It's difficult to understand why, because the language of cinema needs to be renewed." Dod Mantle along with Vinterberg, von Trier, and Dogme 95 reinvigorated the cinematic language by combining the hyper-naturalism of Italian neorealism with the aesthetic characteristics and capabilities of digital video. The effects can be widely seen on movie screens, televisions and today's broad range of viewing devices.

ROGER DEAKINS: GENRE CHAMELEON

Roger Deakins has had quite a different career trajectory than Christopher Doyle or Anthony Dod Mantle. Though all have worked on "Hollywood" films, only Deakins can really be considered a mainstream Hollywood cinematographer, as attested to by his thirteen Academy Award nominations, which is a record for active cinematographers. Since beginning his Hollywood career in 1990 with *Mountains of the Moon* for director Bob Rafaelson, Deakins has worked with a broad array of the industry's most respected directors including Sam Mendes (*Jarhead* [2005], *Revolutionary Road* [2008], *Skyfall* [2012]), Edward Zwick (*Courage Under Fire* [1996], *The Siege* [1998]), Frank Darabont (*The Shawshank Redemption* [1994]), and Martin Scorsese (*Kundun* [1997]), but it is the twelve films he has made thus far with the Coen brothers that most clearly represent his ongoing contributions to the ever evolving cinematic language. Deakins' stylistic innovations are in many ways less immediately visible than Doyle's painterly composition and color palette or Dod Mantle's DV inspired lighting and camera dynamics, but they are no less original nor profound. Not only has Deakins had to apply his talents to the films of over a dozen of Hollywood's leading directors but he has worked in a broad array of genres, periods and styles. With the Coen brothers alone his cinematic range has spanned from the classic noir of *The Man Who Wasn't There* (2001), LA neo-noir of *The Big Lebowski* (1998) and "white" noir of *Fargo* (1996) to the Preston Sturges–inspired cartoon aesthetic comedy of *The Hudsucker Proxy* (1994), the rural musical of *O Brother, Where Art Thou?* (2000), and even the classic Western of *True Grit* (2010). Each of these films has a distinct look, a set of visual characteristics that places it within its generic context, responds to the specific location of setting and shooting, and conveys the frequently idiosyncratic tone of the narrative.

Indeed, when examining the work of Roger Deakins with the Coen brothers one of the first characteristics that leaps from the screen is his extraordinary ability to capture the integral character and inherent thematics of space, particularly in their opening sequences. *The Big Lebowski* begins with a fairly close tracking shot over desert foliage that jibes perfectly with the Texas twang of Sam Elliot's introductory voice-over until the camera picks up a tumbleweed and follows it over a hill to reveal the vast lights of Los Angeles below. The elegant camera movement continues as it tracks the tumbleweed over a freeway, down an LA street and finally to the beach where it rolls toward the water's edge. In this brief sequence Deakins and the Coens have laid out their geography and connected Los Angeles to the tradition of the west, an aspect of that city so often overlooked but certainly significant in this film. *Fargo* begins with what appears to be a grayish-white screen until the faintest image of a bird can be made out in the distance. A set of fence posts appears like a ghostly apparition on the right side of the frame and a pair of soft yellow headlights pierces the whiteness in the distance. By the time the curve of the barren landscape and the slightly more gray asphalt of the road can be discerned, the viewer has been placed within the frozen world, the coldness and emptiness of which provides the setting for the film's noirish (or perhaps *blanc*ish is more correct) story. The sequence of shots showing the sun rising over the Texas desert that opens *No Country for Old Men* (2007) functions in a strikingly similar way. Above the pitch black ground is a small layer cake of faded gold, greenish blue, and darker blue, streaks that get brighter in the next shot which adds a pair of bent telephone poles to the composition, and by the time the sun peaks over the mountains there is no doubt of the unforgiving harshness of the desert landscape. Each new shot is artistically and deliberately composed to continue drawing the spectator in to the place, to visually situate him or her in the world of the story not merely in terms of locale but far more important, in the thematic, dramatic, and emotional meanings of the space.

Connected to Deakins' distinctive handling of space is his understanding of genre conventions and penchant for simultaneously employing them in ways that are recognizable to the viewer while extending them, playing with them, and investing them with a renewed energy and significance. The visual tropes of film noir are extremely well known, so much so that they are often the subject of parody, but the Coens and Deakins in *The Man Who Wasn't There* use many of the familiar lighting and compositional conventions to create the feel of film noir even as these techniques are often inverted. In the introduction to the barbershop where Ed (Billy Bob Thornton) works there are the traditional blinds on the window, only now instead of casting the horizontal shadows so beloved by cinematographers in the 1940s and 1950s, the slashes of light and dark on the window act as bars signifying Ed's entrapment within this not initially threatening space. In the Coens' remake of *True Grit* the landscape is not nearly as lush as in the 1969 original featuring muted greens, a dust colored palette, canted horizons, and most important, a complete abandonment of the pictorial grandeur of the west, the majesty of the open spaces that was such an important element of the western since its earliest days in the first decades of cinema and mythologically enshrined in the work of John Ford. This west is a hard, harsh, and dirty place, no longer operating out of the opposition between garden and wilderness that characterized Ford's films, but rather the same mean streets one might find in any city only here they are unpaved.

Another of Deakins innovative approaches has to do with his handling of that most fundamental cinematic figure, the shot-reverse shot. Conventionally shot-reverse shot, which has the effect of suturing together visual space to produce the illusion of 360-degrees of action without camera, lights, or crew (see Oudart, 1969, Dayan , 1974), is produced either with longer lenses that isolate each individual in the figure or by actually including both subjects in each frame through the use of the over-the-shoulder shot. In his work with Coens Deakins handles this basic shot sequence somewhat differently. Rather than shoot from behind the subject with a longer lens, producing an angle to the shot along with a reduced depth of field, Deakins uses a wide-angle lens (often either a 27mm or 32mm) and places the camera literally between the subjects shooting very nearly straight ahead. The effect here is manifold. First, there is a greater sense of inclusion and intimacy, and Deakins claims from this camera position, "there's a sense of presence, you're right there with somebody [on screen]" (Vimeo interview). This approach also gives them the opportunity to push in to a closer shot from an already isolated single as the move in toward Frank the talkative barber (Michael Badalucco) in *The Man Who Wasn't There* that emphasizes his ever-moving mouth. The qualities of the wide-angle lens add greater force to these movements toward the character. Finally, shooting the close-up with a wider lens allows the image to contain a fair amount of mise-en-scene, so a close up of Marge (Frances McDormand) in *Fargo* as she's in the midst of her investigation maintains her dramatic connection to the snowy waste of the upper Midwest while ace divorce lawyer, Miles Massey (George Clooney) in *Intolerable Cruelty* (2003), is framed at his desk against the open sky of Los Angeles. Though it may seem like a small and simple change in technique, this method of shooting close-ups, particularly in shot-reverse shot contexts, dramatically transforms the spectators, relationship to the material.

One of Deakins most extraordinary innovations in his work with the Coens was in the use of the digital intermediate (DI) in which the film footage is scanned into a digital format so that it can be more broadly and easily manipulated. In the case of *O Brother, Where Art Thou?* Deakins felt that the lush green of the Mississippi terrain would neither translate to film in an aesthetically pleasing manner nor create the correct color palette for the story. When he learned the Coens were planning to shoot in June and July Deakins said "it's going to look like Ireland, not dry and brown and dusty" so he "had to find a way to de-saturate the greens and give the images we were going to shoot the feeling of old, hand-tinted postcards, [which was the look] favored by Joel and Ethan." Working with Kodak's Cinesite in Los Angeles Deakins had the entire film scanned digitally so that its look could be manipulated in post and then printed back out to film for projection. The results are fairly astonishing. The images do indeed have the burnished patina of an old painted postcard, with the greens of the Mississippi landscape transformed into faded golds but the ranges of browns from the trees, the soil, the dirt roads all rendered in their earthy authenticity. The digital color manipulation is also pronounced in the interiors, whether the radio station where the Soggy Bottom Boys record or the Woolworths in which Everett (George Clooney) confronts his rival (Ray McKinnon). The extreme limitation of the color spectrum made possible by the DI never has the often forced or corny feel of a sepia filter but still works to effectively place the film within the dustbowl period. Today of course, almost all films undergo some variation of this digital colorization process, but it was Deakins' attempt to solve a problem posed to him by the Coen brothers that led the way.

The analysis above truly only scratches the surface of Deakins contributions to the art of cinematography through his work with the Coen brothers. Every film that the trio has made together has posed problems and opportunities generated by the broad scope of genres, settings, and tones they are working within, and these challenges have either provoked or allowed Deakins to push the boundaries of cinematography. The sense of geography and genre, color and movement, camera positioning and lens choice are not simply idiosyncratic choices made by these gifted filmmakers but rather represent the same kind of transformation of the language of cinema that Bazin attributed to long-shot long take so many decades ago.

CONCLUSION

Doyle, Dod Mantle, and Deakins have all shot not only in their native and adopted environs but in fact all over the world. Doyle worked in Taiwan and France before ever shooting in Hong Kong, and he and Wong chose to shoot in South America for *In The Mood For Love*. For Dod Mantle's greatest commercial success, *Slumdog Millionaire* he filmed in Mumbai, revisiting the India of his youth, and now moves freely between Denmark, Britain, and the United States even as he continues to live in Copenhagen, Denmark. For *Skyfall* (Sam Mendes, 2014) Deakins shot in China and Turkey as well as the UK, while he made Morrocco stand in for Tibet in *Kundun* (Martin Scorsese, 1997), and shot both *The Kitchen Toto* (Harry Hooks, 1988) and *White Mischief* (Michael Radford, 1987) in Kenya. In his work with the Coen brothers he has taken an extraordinary tour of the United States from the Minnesota of *Fargo* to the Texas and New Mexico of *No Country for Old Men*, from rural Mississippi in *Oh Brother, Where Art Thou?* to urban Chicago in *The Hudsucker Proxy*, places that might as well be different countries. The sense of place and space that comes from this foundation in global mobility informs the work of all three cinematographers. Doyle has said, "So as soon as I see a space, I make decisions. . . . It's basically the Feng Shui. The space implies certain possibilities. The position of elements attracts the body and eye to sit there and avoid that corner. The light falls in a way that implies "peace" or "anxiety" so you respond to that and perhaps enhance it a little. Feng Shui informs intent." Though Doyle is specifically speaking here of specific individual locations within a film, the process works the same way for place in general, and today's transnational cinematographer is more attuned to the cinematic possibilities of place as the boundaries between them and between us continue to dissolve.

To conclude, it is important to recognize that the three cinematographers discussed in this chapter (and the directors with whom they work) are of course not the only ones making significant contributions to the development of the cinematic language. Like the written and spoken languages we use every day, the syntax, grammar, and vocabulary are all changing all the time to the extent that it might be said that every usage of the language reinvents it to some small extent. They are however three of the most prominent figures in the move toward a transnational cinematic expression. In this regard, there are several points of connection between Doyle, Dod Mantle and Deakins. All have explored the range of camera movements made possible by technological developments such as the steadi-cam and techno-crane and the move toward handheld immediacy. They have sought numerous methods of connecting the viewer more intimately to photographed material from Deakins' approach to shot-reverse shot and Doyle's often claustrophobic close-ups to Dod Mantle's positioning his more mobile camera in the middle of the action. Doyle's use of mixed color temperature lighting, Dod Mantle's radically reduced DV-style color palette, and Deakins pioneering use of the DI have all helped change the way color and light are used in the cinema. Ezra and Rowden write, "Transnational cinema imagines its audiences as consisting of viewers who have expectations and types of cinematic literacy that go beyond the desire for and mindlessly appreciative consumption of national narratives that audiences can identify as their 'own.'" (2) The cinematic language that has been pioneered by Christopher Doyle, Anthony Dod Mantle, and Roger Deakins continues its development addressing the needs of a new cinema for a new globalized world.

NOTES

1. Doyle, Christopher, interviewed by Mike Goodridge and Tim Grierson. FilmCraft:Cinematography. Boston: Focal Press, 2011. p. 28.
2. Dod Mantle, Anthony, "Anthony Dod Mantle, DFF/ASC/BSC" Internet Encyclopedia of Cinematographers. Undated. Accessed January 11, 2016.

BIBLIOGRAPHY

Bazin, Andre. *What is Cinema, Vol. 1*. Oakland, CA: University of California Press, 2004.

Burch, Noel. *Life to Those Shadows*. Oakland, CA: University of California Press, 1990.

Deakins, Roger. Interviewed by Tony Zhou. "Joel and Ethan Coen—Shot/Reverse-Shot" February 23, 2016. https://vimeo.com/156455111

Deakins, Roger. Interviewed by Bob Fisher in *American Cinematographer*. October, 2000, pp. 16–22.

Ezra, Elizabeth and Rowden,Terry eds. *Transnational Cinema: The Film Reader*. New York: Routledge, 2006.

Goodridge, Mike and Grierson, Tim. *FilmCraft:Cinematography*. Boston: Focal Press, 2011.

Internet Encyclopedia of Cinematographers. "Anthony Dod Mantlle, DFF/ASC/BSC." Undated. http://www.cinematographers.nl/PaginasDoPh/dod%20mantle.htm

Lieberman, Evan. "Double Vision: Director/Director of Photography Teams and the Development of the Cinematic Language." Paper presented at the University Film and Video Association Conference, Bozeman, Montana, 2014.

Lieberman, Evan and Hegarty, Kerry. "Authors of the Image: Cinematographers Gabriel Figueroa and Gregg Toland." *The Journal of film and Video* 62, nos. 1 and 2 (2010) pp. 31–51.

Metz, Christian. *Film Language: A Semiotics of the Cinema*. Translated by Michael Taylor. Chicago: University of Chicago Press, 1974.

Michelsen, Liselotte and Piil, Morton. "Interview with Anthony Dod Mantle." *Exposure—The Magazine*. Summer, 2003.

Oudart, Jean-Pierre. "Cinema and Suture," in *Cahiers du Cinema: The Politics of Representation,* ed. Nick Browne, Harvard University Press: Mass, 1990.

Rayns, Tony. "Days of Being Wild." Undated. http://www.timeout.com/london/film/days-of-being-wild.

Utterson, Andrew, ed. *Technology and Culture: The Film Reader*. New York: Routledge, 2008, p. 88.

Von Logue Newth, Tom. "Interview: Wong Kar Wei." September 10, 2008. http://screencrave.com/2008-09-10/interview-wong-kar-wai/.

Chapter 5

What Does It Mean to Say that Cinematography Is Like Painting with Light?

Patrick Keating

Cinematographers have been comparing their work to painting for decades.[1] The analogy is intriguing, in part, because it is so imperfect. A painting is made by hand; a film, by a machine. A painting is a single static image; a film unfolds in time and motion. With equal plausibility, one might compare cinematography to writing or sculpture or music—or, to take the most proximate case, to photography. Yet none of these analogies has obtained the currency of the well-worn phrase "painting with light."[2] The phrase may have become something of a cliché, but some cinematographers take the analogy with painting very seriously indeed, proposing that the careful study of painting is essential to a deeper understanding of light and composition. This essay will consider the role that the analogy has played in the theories of three prominent cinematographers: Néstor Almendros, Henri Alekan, and Vittorio Storaro.

In addition to building long careers as world-class cinematographers, Almendros, Alekan, and Storaro developed sophisticated theories about the nature of the craft, articulated in interviews, articles, and in full-length books. A closer look at their statements and writings reveals that each one thinks through the implications of the cinema-painting analogy, while drawing different conclusions. Pointing to the naturalism of his own photography during and after the New Wave, Almendros argues that the task of the cinematographer is to produce pictures that reproduce what the eye sees; he turns to painters like Vermeer because they provide models for the faithful rendering of light's appearances. By contrast, Alekan argues that the primary purpose of studying painting is not depictive but expressive, enhancing the filmmaker's capacity to evoke a wide range of moods. Storaro's extensive writings on the cinema bear some similarities to Alekan's, but the Italian master shifts the emphasis from mood to metaphor. Arguing that writing is just as important a model as painting, Storaro agrees that paintings might inspire individual shots, but he insists that the lighting must change over the course of a film, following an abstract pattern of conflict leading to resolution. Together, these three cinematographer-theorists show that the apparently unanimous endorsement of the "painting with light" analogy masks a complex debate about the value of cinematography, which can be defined in many ways, ranging from the naturalistic to the expressive and the metaphorical.

NÉSTOR ALMENDROS: PAINTING AND THE EYE

Born in Spain and raised in Cuba, Almendros trained at the Centro Sperimentale in Rome. Finding the instruction there to be overly conventional, he moved to New York, where he befriended Maya Deren and Jonas Mekas and wrote a few articles for *Film Culture*. After moving to France, Almendros found success as a cinematographer for Eric Rohmer (*My Night at Maud's*, 1969; *Chloé in the Afternoon*, 1972) and François Truffaut (*The Wild Child*, 1969; *The Story of Adèle H.*, 1975). One of his first American films was Terence Malick's *Days of Heaven* (1978), which won the Academy Award for Best Cinematography, and Almendros spent the last years of his career dividing his time between projects in the United States and in France.[3]

In all his writings, from the early articles in *Film Culture* to the late-career autobiography *A Man with a Camera*, Almendros offers a coherent philosophy of cinematography, rejecting the glamorous formulas of Hollywood in favor of a simpler but no less constructed aesthetic inspired by Neorealism. A 1959 article on "Neorealist Cinematography" lays out this orientation clearly. The first few paragraphs charge that cinematography around the world is suffering under a set of rules "that Berlin had created and Hollywood had imposed"—rules demanding that every scene be illuminated with a key-light, fill-light, backlight, and a background-light.[4] This critique applies to studio hacks and cinematographic masters alike. Almendros professes his admiration for Eugen Schüfftan, Georges Périnal, and Gregg Toland, but he believes that their attempts to transcend convention have produced glaringly artificial works.[5] Commenting on Toland's deep-focus technique, Almendros states, "There is nothing further from human vision than this monstrous mechanical eye with such an optical precision."[6] This criticism rests on a very particular standard for successful cinematography, offering the art an unwavering ideal: not glamour, not expressivity, not technical perfection, but fidelity to human vision.

The rest of the article is a celebration of G.R. Aldo (a.k.a. Aldo Graziati), the Neorealist cinematographer who photographed Luchino Visconti's *La Terra Trema* (1948) and Vittorio de Sica's *Umberto D.* (1952) before he died in an untimely car crash in 1953. As the critique of Toland might suggest, Almendros does not defend Neorealist cinematography on quasi-Bazinian grounds, celebrating the realism of the deep-focus image. Rather, he champions Aldo's willingness to break the commercial cinema's glamour rules ("no more ridiculous 'hair-lights'"), while praising his determination to evoke the appearance of natural daylight.[7] It is the skillful imitation of light's shifting appearances that inspires Almendros's admiration. Significantly, Almendros recognizes that Aldo's lighting effects are constructed, not captured. The earliest Neorealist films, such as *Rome, Open City*, had produced realistic photography by accident, eliminating glamorous touches because of technical constraints. Aldo faced his share of technical challenges, too, but Almendros correctly notes that his films are lit carefully to give the appearance of seeming unlit. For instance, he notes, "The lighting of the actual interiors started from natural light that was coming through the windows and doors, helped by the artificial light, but always through veils or bounced against walls or boards making it much more similar to real daylight."[8] Without supplementary light, it would be difficult to film an interior and exterior at the same time; the difference in light levels is too great for a film stock's latitude. Even when working on location, most Hollywood cinematographers would solve the problem by adding too much frontal light to the interior, thereby destroying the sense that the interior is lit by the sun coming in through the window. Aldo wins Almendros's admiration by adding just enough light to the interior to preserve detail, but not so much as to destroy the sense that the space is lit by the sun.

In this short essay, Almendros twice compares Aldo's work to painting. He notes that the overcast skies of *Cielo sulla Palude* (Augusto Genina, 1949) seem inspired by the paintings of Jean-Francois Millet, and he praises the "oil painting color qualities" of Aldo's only color film, Luchino Visconti's *Senso* (1954).[9] Given Almendros's emphasis on cinematic realism, these references to painting might seem contradictory. Why not aspire toward photographic objectivity instead? But Almendros's version of realism values iconic resemblance more than indexicality. A skillful painter is attentive to the subtle gradations of light, providing a model for the sensitive cinematographer to follow.

When Almendros himself became a cinematographer, the study of painting proved to be a consistent source of inspiration. His autobiography puts it plainly: "When I am preparing for a film I generally think of a painter or a school of painting."[10] Painters are categorized according to their ability to render certain types of light: "Vermeer for daytime interiors and La Tour for night interiors that are lighted by the luminous source of a flame. To these I must add Rembrandt and Caravaggio for effects of chiaroscuro, and Manet, Renoir, and other Impressionists for day exteriors."[11] Painters are privileged not because of their high cultural status, but because they teach the cinematographer to look at the world. Almendros is equally impressed with the works of silent-era filmmakers like Griffith and Feuillade, who made films on sunlit stages before the three-point lighting system came to dominate commercial cinema in the late 1910s.[12] Vermeer and Griffith, Renoir and Feuillade—all offer inspiration by attending to the nuances of light and shade as they fall upon a figure.[13] All offer alternatives to the artificiality of Hollywood's three-point lighting system.

In practice, Almendros's philosophy privileges the use of available light. "I like to work with limited means," he explains, "because I am looking, above all, for the result on the film to be as close as possible to what the human eye sees."[14] However, Almendros recognizes that the eye and the film stock do not always "see" the same way, especially when faced with "contre-jour" lighting, directed toward the lens. "The human eye sees contre-jour; the film stock does not. So, a little bit of supplementary light will give what the eye receives." For instance, Truffaut's *The Wild Child* features several scenes with characters standing in front of open windows and doorways. In Figure 5.1, the staging contrasts the two men playing backgammon in the foreground with the boy

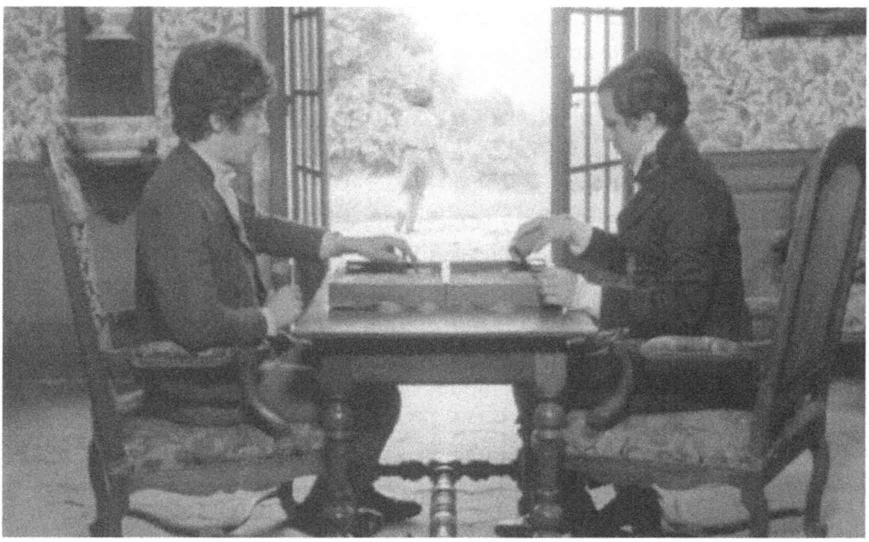

Figure 5.1 In *The Wild Child,* Almendros creates the impression of a room lit by soft daylight from the window.

walking in nature in the background. As in the films of Aldo, natural sunlight flows through the window. This light bounces off the floor, providing soft light throughout the space. Still, the room would look too dark without careful work on Almendros's part. After adding a fill light from above (notice the shadow underneath the tabletop), he sets his exposure for the interior and allows the exterior to overexpose slightly, thereby creating the impression of seeing a bright daytime room that is still somewhat dimmer than the brilliant world beyond.

Later in his career, Almendros worried that his colleagues in the New Wave had turned the innovation of soft lighting into the cliché of flat lighting: "That shadowless light always shining down from a strange sky (the ceiling), by day or night, had eventually destroyed visual atmosphere in modern cinema."[15] Eager to reject Hollywood's conventions, New Wave cinematographers imitating Raoul Coutard created a new convention, bouncing light off the ceiling in every scene. For lighting historian Fabrice Revault d'Allonnes, the strategy marked Coutard and his followers as truly modern, willing to acknowledge the simple fact that light in the real world is often meaningless, utterly indifferent to human concerns like narrative.[16] Almendros was more skeptical of the flat-lighting aesthetic, worrying that post-New Wave practitioners had forgotten the core principle of cinematography—the principle that the cinematographer should always strive to reproduce "what the eye would really see."[17] Paintings, offering everything from the shadows of Rembrandt to the sunlight of Monet, provide a valuable corrective, representing the full variety of lit spaces to counter the predictable three-point lighting schemes of Hollywood, on the one hand, and the equally predictable flatness of post-New Wave cinematography, on the other.

HENRI ALEKAN: PAINTING AND EXPRESSION

Alekan's career in cinematography spanned nearly half a century. In 1946, he accomplished the remarkable feat of photographing two films in radically different styles: the highly pictorial *The Beauty and the Beast*, directed by Jean Cocteau, and the documentary-style *The Battle of the Rails*, directed by René Clement. Though he continued to work professionally for several decades, he never collaborated with any of the major figures of the New Wave, instead photographing glossy international productions like *Austerlitz* (Abel Gance, 1959) and *Topkapi* (Jules Dassin, 1964). In the 1980s, his reputation among cineastes enjoyed a major revival, on the strength of his collaborations with several noted auteurs: Raoul Ruiz (*On Top of the Whale*, 1982), Amos Gitai (*Esther*, 1986), and, most famously, Wim Wenders (*Wings of Desire*, 1987).

Meanwhile, Alekan produced a major work on the theory of cinematography: *Des lumières et des ombres*, first published in 1979.[18]

With its coffee-table-book size, *Des lumières et des ombres* is a remarkable work, featuring lengthy chapters on the topics of sunlight, nocturnal light, and artificial light. It is not a how-to manual—that is, Alekan does not instruct readers on how to plug in a lamp or how to attach a silk to a C-stand. Instead, he starts with a precise description of what the ordinary observer might see when looking at each sort of light. For instance, on the subject of moonlight, he writes, "The lunar orbit through space is accomplished like that of the sun, and, for the observers that we are, from East to West, with, as a consequence, a variable angularity that creates, as with sunlight, cast shadows of changing density according to whether the atmosphere is cloudy or not."[19] Alekan asks the cinematographer to attend to details such as this. At any time of day, what is the source of light? At what angle does it fall? How dense are the resulting shadows?

Part of the job of the cinematographer is to learn how to imitate the wide variety of lighting effects that are observable in everyday life. For instance, after noting that human vision does not register strong colors under the dim illumination of moonlight, Alekan advises that "the filmmaker therefore reduces the palette by means of artificial techniques to give the appearance of moonlight."[20] Like Almendros, Alekan starts with the close observation of natural light. Like Almendros, he advises the cinematographer to produce images that resemble what a human observer might see under familiar lighting conditions, such as moonlight. Unlike Almendros, Alekan refuses to privilege naturalistic effects, arguing that fidelity is less important than expression. Filmmakers and painters do not represent moonlight in all its nuances because they care about accurate representation, but because they hope to use moonlight (and all the other lighting effects) to shape the viewer's emotional experiences.

Before turning to Alekan's critique of naturalism, it is useful to ask: How does lighting express an emotion? Alekan's answer: by appealing to a spectator's memory. In his chapter on sunlight, Alekan writes, "The spectator in front of a painterly or filmic work is placed in a condition of receptivity by the grace of his faculty of memorization, the plastic information before him revealing sensations and feelings pre-registered through his contact with nature."[21] As spectators, each of us has spent a lifetime observing the sun rise and set, and so each has built up a lifetime's worth of memories about sunlight. We have experienced the sun at noon, the sun piercing through the trees, and the sun disappearing behind the clouds. When a painter or filmmaker decides to represent a scene that takes place five minutes after sunset, the decision calls forth a flood of memories in the spectator's mind. For example, Alekan suggests that moonlight provides a powerful contrast to artificial light when they appear together in the same frame: "The result is a battle between these opposing forces: moonlight and artificial light, the one symbolizing continuity, serenity, simplicity, melancholy, poetry, coldness, death; and the other, vivacity, complexity, instability, materiality, warmth, life."[22] The cinematographer manipulates thoughts to get at the associated emotions.

At this point, it seems fair to charge that Alekan may be exaggerating the homogeneity of human memory. Different people have different experiences of moonlight, and indeed a single person might have many conflicting memories. Though the book as a whole tends toward universalizing claims, there are passages in which Alekan suggests that his claims are reserved for spectators engaging with a particular art historical tradition—that of Western painting.[23] This tradition has trained viewers how to respond to light, and he hopes that the cinema will extend its conventions. Continuing his discussion of the evocative contrast between moonlight and artificial light, Alekan points to Van Gogh's *Café Terrace at Night* (1888) and Magritte's *The Empire of Lights* (1953) as models. Studying such paintings will help the cinematographer gain command of their complex lighting effects, producing a wider range of emotional effects.

Early in his career, Alekan had worked as an assistant and operator for Eugen Schüfftan, the German cinematographer who photographed several classics of Poetic Realism in the 1930s, most notably *Port of Shadows* (Marcel Carné, 1938). As we have seen, Almendros admired Schüfftan, but held reservations about his overly artificial style. Alekan shares no such reservations. In a written interview, Alekan enthuses about his old boss: it was Schüfftan, he says, "who initiated me into the secrets of his art, which held two principles: observation of the composition and the lighting of the Old Masters of painting; transposition and application to the cinema of the rules revealed by paintings."[24] Studying paintings might teach a cinematographer how to imitate sunlight, moonlight, or artificial light, but imitation is not the ultimate goal. Alekan quotes Schüfftan as saying, "Look here, I'm not doing a naturalistic lighting; I'm doing lighting as I feel it, emotional lighting."[25] If Schüfftan learned any lesson from painting, it was the lesson that the image can express emotions.

This willingness to distinguish emotional lighting from naturalistic lighting differentiates Alekan from Almendros. To be sure, Almendros shares Alekan's belief that visual style can and should express emotions. The lighting in Rohmer's *My Night at Maud's* evokes the winter chill of Clermont-Ferrand, and the resulting images neatly express the protagonist's feelings of repression. But Almendros voluntarily restricts his options by committing to work in a more muted style, in which every shot aspires to the ideal of what the eye would see. By contrast, Alekan insists that naturalism is one stylistic option among many. In interviews, he is particularly proud of the fact that he photographed *The Beauty and the Beast* and *The Battle of the Rails* in the same year. The former film was a romantic fantasy, and Alekan modeled his dreamlike images on the engravings of Gustave Doré. The latter film was based on real events, and Alekan's style employed a self-effacing style with the roughness of a documentary.[26] Rather than duplicate the lighting of the world over and over again, Alekan prefers "to invent a style for every film."[27]

Coincidentally, the still photographer on *The Beauty and the Beast* was none other than G.R. Aldo, who soon returned to Italy to photograph the Neorealist classics that Almendros admired so much. When Alekan turns to the subject of Neorealism in his book, he is at pains to point out that the best Neorealist photography is carefully composed, with considerable thought given to the choice of lighting effects.[28] Even the decision to rely on available light is a creative decision, requiring great care in the selection of available light to use. Whereas some champions of Neorealism, such as Siegfried Kracauer, celebrate the cinema's affinity for moments of chance, Alekan rejects the accidental as a source of the aesthetic.[29] "The pictorial art, like the cinematic art, is not born of chance," he writes firmly. "It is willfully constructive."[30] By nominating painting rather than photography as the true predecessor of cinematography, Alekan puts forth a definition of the cinematographer as a figure who controls each composition down to the finest detail, no matter how much the composition is based on the observation of striking lighting effects in the real world.

If Alekan is willing to accept the achievements of Neorealist cinematography, he is much more ambivalent about the imagery of the French New Wave. On the one hand, he recognizes the value of the decision to leave the studios and to work on real locations, especially since studio lighting had grown so conventional. On the other hand, he amplifies Almendros's charge that too many cinematographers solved the resulting practical problems by relying on bounced lighting in every situation, thereby sacrificing the variable lighting effects that mark the best cinematography.[31] Describing the technique as "lumière anglaise," Alekan suggests that New Wave cinematographers blindly copied the flat look of English fashion photography.[32] The reference to fashion photography might suggest that the technique was overly stylized, but in fact Alekan's point is that New Wave cinematography was excessively naturalistic, adopting a plain affect as a way of signaling the rejection of studio aesthetics. "My view," he said in an interview, "is that you have to break the banality of naturalism. We get naturalism in our everyday lives. Artists are made to invent something else."[33] The cinematographer is not an observer, but a creator.

Soon after he published the first edition of his book, Alekan enjoyed a creative revival, as directors like Wenders, Alain Robbe-Grillet, and Joseph Losey began requesting his services. Some of the late films rely on sharp

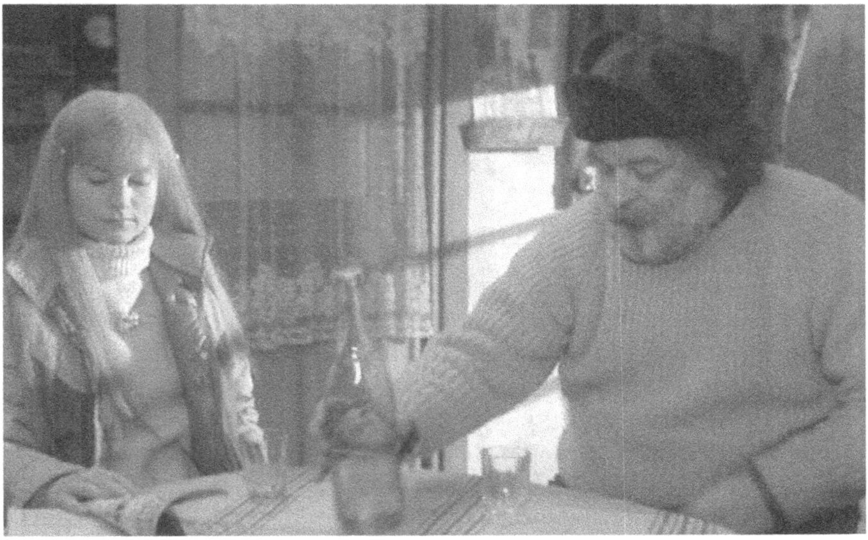

Figure 5.2 In *La Truite*, Alekan uses a cool, hard window light for expressive effect.

juxtapositions, as in the contrasts between color and black-and-white footage in *Wings of Desire*, or between the sunlit daytime scenes and the artificial, colorfully gelled nighttime scenes of *Esther*. Other films feature more understated interventions. In Figure 5.2, from *La Truite* (Losey, 1982), the lighting initially suggests the familiar effect of sunlight flowing through the window. However, Alekan makes two adjustments for expressive purposes. First, he employs a hard light to mimic the sunlight, an unexpected choice given that the French exteriors are characterized as overcast throughout the film. This decision to violate naturalistic expectations allows Alekan to establish an expressive visual opposition between the two characters: Isabelle Huppert's vulnerable-yet-powerful protagonist Frédérique is side-lit, while Jean-Paul Roussillon's pathetic old lecher sits in moderate darkness. Second, Alekan has added a greenish blue gel to the outside lamp, to better suit the film's bleak mood. This is not the warm sun flooding the room, but a light that is ugly and cold.

For Almendros, painting may provide a useful starting point, but the ultimate objective of the cinematographer is to produce images that resemble what the eye might see. For Alekan, painting establishes an aspirational model. Only the most determined cinematographer can produce images that are worthy of being classed as contributions to the pictorial tradition. Vittorio Storaro will offer a third way of thinking about the cinema's relationship with painting, situating the analogy within a broader theory in which cinematography is best conceived as a form of writing.

VITTORIO STORARO: PAINTING AND WRITING

During the 1970s, Vittorio Storaro gained worldwide fame for his work with Bernardo Bertolucci on such projects as *The Conformist* (1970), *The Last Tango in Paris* (1972), and *1900* (1976). The following decade, Storaro won the Academy Award for Best Cinematography three times, while working with three different directors: on Francis Ford Coppola's *Apocalypse Now* (1979), Warren Beatty's *Reds* (1981), and Bertolucci's *The Last Emperor* (1987). Offscreen, Storaro developed a reputation as a thoughtful, if idiosyncratic, advocate for the art of cinematography. *American Cinematographer* published a number of articles on his work, later collected in the anthology *Writer of Light*. In 2003, Storaro published an ambitious three-volume work in Italian and English, *Scrivere con la Luce*, or *Writing with Light*.

Structurally, Storaro's opus resembles Alekan's *Des lumières et des ombres*, moving one by one through the various forms of lighting we encounter in life and in film, featuring numerous cinematic and painterly examples along the way. For instance, the first half of the first volume, *The Light*, offers chapters on the following topics: shadow, penumbra, light, punctiform light, multiform light, natural light, artificial light, the sun, the moon, and the infinite. In each chapter, a brief text written by Storaro is set alongside illustrations of two famous paintings, followed by several pages of frame enlargements from Storaro's own films. The pattern repeats itself several times: text alongside paintings, followed by films. Georges de la Tour accompanies Storaro's comments on "shadow," while Vermeer appears in the section on "penumbra." For Angela Dalle Vacche, the books propose a "position statement about cinematography's reliance on painting."[34]

Storaro's words rarely comment on the paintings directly, but the layout asks the reader to look at the paintings with Storaro's evocations in mind. In the section on "punctiform light" (that is, hard light produced by a single-point source), Storaro writes, "Its linearity in space has permitted the clear-cut separation between the two magnetic forces of Light and Shade, thus implying the division between two worlds: Good and Evil; Male and Female; Conscious and Unconscious, thus creating a natural conflict in motion that was the origin of everything."[35] Storaro's descriptions are less precise than Alekan's, but even more evocative, borrowing the language of psychoanalysis to suggest how an image might appeal to the unconscious mind. One of the accompanying illustrations is Caravaggio's *The Conversion of St. Paul* (1600). With Storaro's text in mind, we might see the sharp contrasts between light and dark as a symbol for the abruptness of Paul's conversion from one state to another. Notice that the painting represents a narrative event—a moment of change. Storaro asks us to consider how lighting helps to "write" the story—not by highlighting important details, in the manner of a conventional Hollywood film, but by symbolizing the abstract idea of conflict itself.

Following the text, each chapter offers several pages of frame enlargements from Storaro's own films, juxtaposed with quotations attributed to artists and philosophers from around the world, such as Plato, Averroës, and Goethe. The section on punctiform light includes an illustration from *The Conformist*, as shown in Figure 5.3. As the repressed Clerici (Jean-Louis Trintignant) sits on the couch, a lamp beaming through Venetian blinds casts sharp lines across the wall. The placement of the lamp is carefully judged: Trintignant holds his face so that his right eye is just on the border between light and shadow, a suitable symbol for a character who remains trapped in ignorance even though he is so close to wisdom. The film explicitly solicits such metaphorical readings when the character of the professor recounts the myth of Plato's cave. Whereas Almendros claims that his primary interest is capturing (or imitating) the appearance of light, Storaro privileges the power of lighting to add layers of symbolic commentary to the work.[36]

A few years after he published the first edition of his books, Storaro photographed the biographical film *Caravaggio* (Angelo Longoni, 2007). In a subsequent interview, Storaro told interviewer Bob Fisher about his initial encounter with Caravaggio's painting *The Calling of St. Matthew* (1600): "There is a beam of light that goes from the top to the bottom of the painting, dividing it into two parts. One side is in daylight and the other side is in

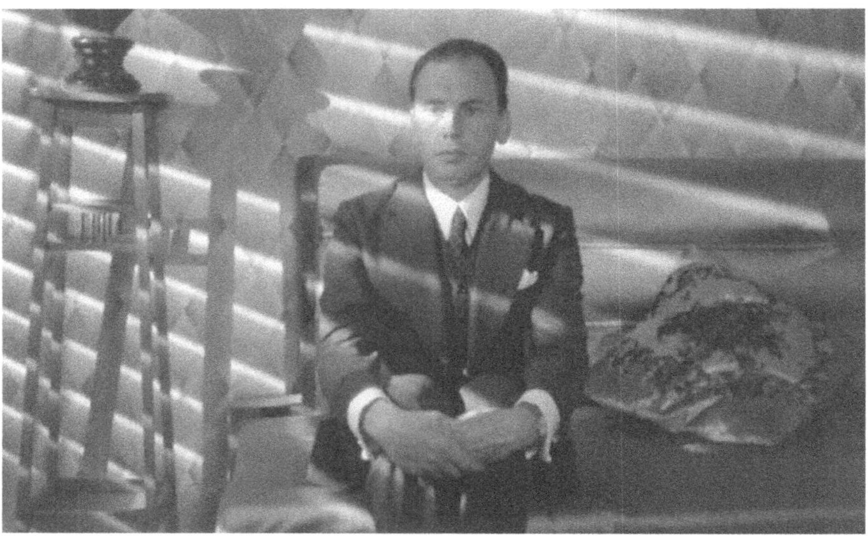

Figure 5.3 In *The Conformist,* a Venetian blind effect emphasizes the symbolic opposition between light and dark.

darkness. I recall thinking they represented the human and the divine sides of life and our unconscious and conscious beings. That was the first time that I saw light and darkness used as metaphors for life and death." Significantly, Storaro goes on to compare the painting to a work of literature: "I also remember reading a book by William Faulkner called *Absalom, Absalom!*, where one of the main characters explains how a beam of sunlight penetrated and divided a room like it was separating periods in another character's life. It was the same concept as *The Calling of Saint Matthew*."[37] Whereas Alekan had worked hard to present cinematography as the continuation of a tradition of painting, Storaro acknowledges the influence of painting only to slide into another analogy: cinematography as writing. Light has the power to evoke ideas. Anything the writer can say with words, the cinematographer can say with light.

With this invocation of writing as a model, we might suspect that Storaro is making a familiar rhetorical move: by championing the artificiality of his craft, he grants the cinematographer increased status as a creator. But Angela Dalle Vacche reminds us that Storaro's work, however stylized it may seem, shares an important affinity with Neorealism as Bazin conceived it. Celebrating the chemical reaction that occurs when light hits the film stock, Storaro defines the cinema as a record of "the real presence of objects and bodies in front of the camera."[38] The cinematographer writes *with* light, but remains humble before the mysterious power of light itself.

Writing provides an important point of comparison because it introduces the theme of time. Rejecting the label "director of photography," Storaro reminds us that cinematography is, linguistically, the writing of movement. He tells one interviewer, "Photography is still just ONE single image. Cinematography contains movement. Not only different images, but a kind of SENTENCE. Cinematography is an art form because it is WRITING the story using the grammar of light, shadows, and color—like a writer with words."[39] Following the example of writing, the cinematographer must develop the ability to unfold an expression over time. This proposal contributes to a long-standing debate that stretches back at least as far as Lessing's *Laocoön*, which used the contrast between space and time to ponder the relative powers of painting and poetry.[40] The cinema's capacity for movement distinguishes the art from photography—and, by extension, from painting, which is also restricted to the single image.

Of course, Storaro is not the first cinematographer to suggest that the job entails telling a story. Hollywood cinematographers had been championing the value of storytelling since the studio days.[41] But Storaro's ideas about what counts as storytelling are unusually abstract. "What the cinematographer does," he explains, "is to write the story with a beginning, an evolution, and an end."[42] In practice, this means that the cinematographer identifies a crucial conflict in the story and represents that conflict by establishing an opposition between different forms of light: for instance, between two colors. Then comes the crucial step: the opposition should not remain static over the course of the film. Rather, the opposition should develop over time, culminating at the end of the film with imagery that reconciles the conflict. In Storaro's words, "To me, making a film is like resolving conflicts between light and dark, cold and warmth, blue and orange or other contrasting colors."[43] Storaro turns each narrative into an abstraction—conflict leading to a resolution—and uses this abstract scheme to guide the film's visual progression.

In interviews and articles, Storaro has outlined the contrasts that structure many of his films. In *Apocalypse Now*, the conflict between Vietnam and the United States is rendered as a conflict between natural lighting and overwhelming artificial lighting, developing the theme of "imposing one culture on another."[44] In *Life without Zoe* (Coppola, 1989), the conflict between the father and the mother becomes a conflict between the sun and the moon.[45] Whereas the despairing *Apocalypse Now* progresses toward the color black, associated with Marlon Brando's Kurtz, the more hopeful *Life without Zoe* culminates in a scene depicting the sun and moon at the same time, suggesting a possibility of harmony within the family. Other films move beyond the binary opposition to develop a more complicated progression. In *Little Buddha* (Bertolucci, 1993), the three children are identified with the three primary colors of the additive system: red, green, and blue, symbolizing the elements of fire, water, and air, respectively. The film builds toward white light, the synthesis of red, green, and blue.[46] Most remarkably, for *The Last Emperor*, Storaro devised a progression based on the seven colors of the spectrum. As the protagonist grows to maturity, the film moves through each color, in order: red, orange, yellow, green, blue, indigo, and violet, eventually arriving at white light, figuring the theme of synthesis.[47]

By the latter half of his career, Storaro had come to believe that each color carries specific symbolic meanings. In *Writing with Light*, he explains, "RED is an expression of the life force; ORANGE is a passionate embrace; YELLOW indicates intuition; GREEN is the symbol of inner life; BLUE represents spiritual energy; INDIGO communicates material satisfaction; VIOLET is sacred, the color of introspection."[48] These associations seem arbitrary, but Storaro insists that they are widely shared, the result of unconscious associations people make as they grow from childhood to adulthood. His book's wide range of quotations—Greek philosopher Plato, Italian scientist Galileo, Japanese novelist Junichiro Tanizaki—gestures toward this ideal of universality. However, it must be said that this universalizing tendency is probably the weakest part of Storaro's theory. Some of Storaro's oppositions are based on troubling ideological assumptions, as when he maps the culture/nature opposition onto the colonial power and the colonized in *Apocalypse*

Now, or when he conflates the female and the maternal to justify the use of orange in *The Last Emperor*. By refusing the acknowledge the arbitrariness of his oppositions, Storaro weakens the analogy with language that he has worked so hard to construct.

When the *American Cinematographer* journalist Bob Fisher sensibly asks Storaro to address the fact that different cultures might interpret colors in different ways, Storaro insists, "Each color has a specific wavelength of energy, which we perceive the same way that we feel vibrations. Even if they aren't consciously aware of it, the audience can feel a difference between high and low wavelengths of energy."[49] Lighting has a direct impact on the spectator's body. Red increases the blood pressure; blue and green lead the nervous system to a decrease in function.[50] Remarkably, Storaro suggests that colors produce these effects whether we see them or not. "The human body doesn't see with only two eyes. It sees with the entire body. There is an energy that arrives at our body that gives a kind of impression to which we will react emotionally."[51] Even with our eyes closed, he argues, our bodies feel the differences between the wavelengths. Turning from psychology to physiology, Storaro has pushed the argument to the point that the analogy with language breaks down. His grammar is not a set of conventions to be learned, but a set of triggers that the cinematographer uses to manipulate the spectator's bodily experiences.[52]

The real importance of the writing analogy is not to position cinematography as a semiotic system, built on the manipulation of conventions. Rather, it is to emphasize the way lighting must change from the beginning of the film to the end. Like Almendros and Alekan, Storaro believes that cinematography starts with the careful observation of light in the everyday world. But this careful observation of light discloses cinematography's true power—the power to evoke conflicts that move toward a resolution over time.

CONCLUSION

The analogy between cinematography and painting persists to this day, even after the widespread adoption of digital tools. In a recent article on digital color grading, editor "Ace" Gates concludes with the familiar metaphor: "The pigments are pixels and the brushes are digital when painting the video canvas."[53] More critically, cinematographer John Bailey describes the digital intermediate process as a "second palette" for cinematographers—though he worries that color graders may use this palette in problematic ways.[54] Both the defenders and the critics of digital technology appeal to the painting analogy in order to situate the new tools within a larger history—not necessarily the history of Old Masters painting, but the history of pre-digital cinematography, which turned the phrase "painting with light" into a commonplace.

Though the analogy with painting can sometimes seem like an empty cliché, it serves to express underlying assumptions about the craft of cinematography. In the abovementioned article on digital color, Gates draws a distinction between color correction and color grading. The former seeks "an accurate portrayal of the subject as it would be viewed from the human eye"; the latter accomplishes the more ambitious task of "altering an image for aesthetic and communicative purposes."[55] Because color grading furthers the story by developing mood, tone, and theme, the craft can be considered "an art form all its own."[56] Celebrating the power of color grading, Gates aligns himself with the thinking of Alekan and Storaro, against the more naturalistic ambitions of Almendros's theory. Gates is not alone—much of the rhetoric about color correction emphasizes the ideals of expression, whether making the analogy to painting explicit or not. Cinematographer Roger Deakins argues that the cinematographer should take advantage of any tool that might help one "develop as a digital storyteller."[57] Colorist Stefan Sonnenfeld defends the job by noting, "The biggest thing is that we can work with [cinematographers] to create looks that set the tone of stories."[58] Similarly, colorist Jill Bogdanowicz has explained how she created different looks for each time period in *The Grand Budapest Hotel* (Wes Anderson, 2014), "pushing and pulling the color in different directions."[59] Like Storaro, Bogdanowicz enhances the story by developing a progression of looks that change over time. Significantly, the system she uses to do the pushing and pulling is named after a painter: Da Vinci.

For Almendros, Alekan, and Storaro, the art of cinematography starts with the careful observation of different kinds of light, noticing the differences between dull sunlight at the start of an overcast day and brilliant sunlight in the middle of a hot afternoon. In the digital age, the analogy with painting lingers on, but with a shifting set of connotations, placing ever more emphasis on the power of filmmakers—directors, cinematographers, and colorists—to reshape the image at will.

NOTES

1. An early example of the idea appears in a 1927 issue of *The Moving Picture World*. A profile of cinematographer Joseph La Shelle compares the camera to "the brush and easel of the portrait artist." See "An Artist in Camera Angles," *The Moving Picture World*, April 23, 1927, 714.

2. Two oft-cited works bearing this title are Victor Milner, "Painting with Light," in *The Cinematographic Annual*, vol. 1, ed. Hal Hall (Los Angeles: American Society

of Cinematographers, 1930): 91–108; and John Alton, *Painting with Light* (1949; Berkeley: University of California Press, 1995).

3. For more details on Almendros's career, see his autobiography, *A Man with a Camera*, trans. Rachel Phillips Belasch (New York: Farrar, Straus, and Giroux, 1984).

4. Néstor Almendros, "Neorealist Cinematography," *Film Culture* 20 (1959): 39.

5. Almendros, "Neorealist Cinematography," 39.

6. Almendros, "Neorealist Cinematography," 40.

7. Almendros, "Neorealist Cinematography," 41.

8. Almendros, "Neorealist Cinematography," 41.

9. Almendros, "Neorealist Cinematography," 41–42.

10. Almendros, *A Man with a Camera*, 102.

11. Almendros, *A Man with a Camera*, 239; Almendros also discusses Vermeer in Sharon Russell, *Semiotics and Lighting: A Study of Six Modern French Cameramen* (Ann Arbor: UMI Research Press, 1981), 92.

12. For references to Griffith and Feuillade, see Almendros, *A Man with a Camera*, 85; and René Predal, "Les grands operateurs (III)," *Cinema 73,* 172 (January 1973): 99.

13. Karl Brown, who worked as Billy Bitzer's assistant on several films by D.W. Griffith, claims that he, too, was inspired by Vermeer. See Brown, *Adventures with D.W. Griffith*, ed. Kevin Brownlow (New York: Da Capo Press, 1976), 51.

14. Almendros, quoted in Gerard Langlois, "Les montreurs d'ombres VIII.Néstor Almendros: L'homme aux miroirs," *Lettres françaises* 1352 (September 23–29, 1970): 18. My translation.

15. Almendros, *A Man with a Camera*, 7.

16. Fabrice Revault d'Allonnes, *La lumière au cinéma* (Paris: Cahiers du cinéma, 1991), 9–10.

17. Almendros, *A Man with a Camera*, 9.

18. For a useful summary of Alekan's career, see Benjamin Bergery, "Henri Alekan: The Doyen of French Cinematography," *American Cinematographer* 77, no. 3 (March 1996): 46–52.

19. Henri Alekan, *Des lumières et des ombres* (Paris: Le Sycamore, 1984), 97. My translation.

20. Alekan, *Des lumières et des ombres*, 97. My translation.

21. Alekan, *Des lumières et des ombres*, 36. My translation.

22. Alekan, *Des lumières et des ombres*, 99. My translation.

23. See footnote 2 in Alekan, *Des lumières et des ombres*, 15.

24. Alekan, quoted in René Predal, "Les grands operateurs (IV)," *Cinema 73* 173 (February 1973): 93. My translation.

25. While speaking with Benjamin Bergery, Alekan is recalling a conversation he had with Schüfftan decades before. See Bergery, "Henri Alekan," 47.

26. Alekan, quoted in Predal, "Les grands operateurs (IV)," 94.

27. Alekan, quoted in Bergery, "Henri Alekan," 50.

28. Alekan, *Des lumières et des ombres*, 243.

29. In *Theory of Film*, Kracauer celebrates "the affinity of film for haphazard contingencies." Later in the same chapter, he praises *Bicycle Thieves* and *Umberto D.* for their rendering of the "flow of life." See Siegfried Kracauer, *Theory of Film: The*

Redemption of Physical Reality (1960; Princeton: Princeton University Press, 1997), 62, 73.

30. Alekan, *Des lumières et des ombres*, 243. My translation.

31. Alekan, *Des lumières et des ombres*, 246.

32. Bergery, "Henri Alekan," 51.

33. Alekan, quoted in Bergery, "Henri Alekan," 51.

34. Angela Dalle Vacche, "Chiaroscuro: Caravaggio, Bazin, Storaro," *Senses of Cinema* 53 (December 2009), *Senses of Cinema* website, accessed January 9, 2016, http://sensesofcinema.com/2009/feature-articles/chiaroscuro-caravaggio-bazin-storaro.

35. Vittorio Storaro, *Writing with Light, Part One: The Light* (New York: Aperture, 2003), 52.

36. This distinction between Almendros and Storaro applies to their written theories more than it does to their films. In practice, Almendros's lighting may also appeal to symbolism, as in the example from *The Wild Child*, where it is no accident that the "uncivilized" boy is associated with bright, uncontrollable sunlight.

37. Storaro, quoted in Bob Fisher, "Vittorio Storaro Paints with his Camera," *Moviemaker* website, accessed January 9, 2016, http://www.moviemaker.com/articles-moviemaking/vittorio-storaro-cinematographer-caravaggio-20071127. Originally published in January, 2008.

38. Dalle Vacche, "Chiaroscuro."

39. Storaro, quoted in Ray Zone, "The Literature of Light: An Interview with Vittorio Storaro, ASC, AIC," in *Writer of Light: The Cinematography of Vittorio Storaro, ASC, AIC*, ed. Ray Zone (Hollywood: ASC Press, 2000), 80.

40. See Gotthold Lessing, *Laocoön: An Essay on the Limits of Painting and Poetry*, trans. Edward Allen McCormick (Baltimore: The Johns Hopkins University Press, 1984), 78. Dalle Vacche contrasts Storaro's conclusions with Lessing's in "Chiaroscuro."

41. For instance, longtime Twentieth Century-Fox cinematographer Arthur Miller, once said, "A good picture, as we all know, starts with a story. The next thing is to tell that story pictorially." Miller, quoted in Leonard Maltin, *The Art of the Cinematographer: A Survey and Interviews with Five Masters* (New York: Dover Publications, 1978), 64.

42. Storaro, quoted in Zone, "The Literature of Light," 80.

43. Storaro, quoted in "Q&A with Vittorio Storaro, ASC, AIC, on the Filming of *Apocalypse Now*," in *Writer of Light: The Cinematography of Vittorio Storaro, ASC, AIC*, ed. Ray Zone (Hollywood: ASC Press, 2000), 8.

44. Storaro, quoted in "Q&A with Vittorio Storaro," 8.

45. David Heuring and Nora Lee, "*Life without Zoe*," in *Writer of Light*, 41–42.

46. Bob Fisher, "Large Format Expands *Little Buddha*," in *Writer of Light*, 53.

47. Storaro, quoted in Zone, "The Literature of Light," 85–86.

48. Storaro, *Writing with Light, Part One: The Light*, 16.

49. Storaro, quoted in Bob Fisher, "Master of Light and Motion," in *Writer of Light*, 70.

50. Storaro, quoted in "Mating Film with Video for *One from the Heart*," in *Writer of Light*, 29.

51. Storaro, quoted in Zone, "The Literature of Light," 82.

52. After stating, "I believe that we have a kind of grammar," Storaro argues, "Each one has its own meaning. Each one has a specific emotion." Storaro, quoted in Zone, "The Literature of Light," 83.

53. Chris "Ace" Gates, "Color Grading," *Videomaker* 28, no. 3 (September 2013): 56.

54. John Bailey, "The DI Dilemma, or: Why I Still Love Celluloid," *American Cinematographer* 89, no. 6 (June 2008): 93.

55. Gates, "Color Grading," 54.

56. Gates, "Color Grading," 53.

57. Roger Deakins, "The DI, Luddites and other Musings," *American Cinematographer* 89, no. 10 (October 2008): 79.

58. Bob Fisher, "The Big Timers: Meet the DI Colorists," *Studio Daily* website, accessed January 7, 2015, http://www.studiodaily.com/2005/06/the-big-timers-meet-the-di-colorists. Originally published June 1, 2005.

59. "Attaining the Wes Anderson Aesthetic: Fine-Tuning the Fantasy of *The Grand Budapest Hotel*," *Digital Video* 22, no. 5 (May 2014): 53.

BIBLIOGRAPHY

Alekan, Henri. *Des lumières et des ombres*. Paris: Le Sycamore, 1984.

Almendros, Néstor. *A Man with a Camera*. Translated by Rachel Phillips Belasch. New York: Farrar, Straus, and Giroux, 1984.

———. "Neorealist Cinematography." *Film Culture* 20 (1959): 39–43.

Alton, John. *Painting with Light*. Berkeley: University of California Press, 1995.

"Attaining the Wes Anderson Aesthetic: Fine-Tuning the Fantasy of *The Grand Budapest Hotel*." *Digital Video* 22, no. 5 (May 2014): 53–54.

Bailey, John. "The DI Dilemma, or: Why I Still Love Celluloid." *American Cinematographer* 89, no. 6 (June 2008): 92–97.

Bergery, Benjamin. "Henri Alekan: The Doyen of French Cinematography." *American Cinematographer* 77, no. 3 (March 1996): 46–52.

Brown, Karl. *Adventures with D.W. Griffith*. Edited by Kevin Brownlow. New York: Da Capo Press, 1976.

Dalle Vacche, Angela. "Chiaroscuro: Caravaggio, Bazin, Storaro." *Senses of Cinema* 53 (December 2009). http://sensesofcinema.com/2009/feature-articles/chiaroscuro-caravaggio-bazin-storaro.

Deakins, Roger. "The DI, Luddites and Other Musings." *American Cinematographer* 89, no. 10 (October 2008): 78–83.

Fisher, Bob. "The Big Timers: Meet the DI Colorists." *Studio Daily*, June 1, 2005. http://www.studiodaily.com/2005/06/the-big-timers-meet-the-di-colorists.

———. "Vittorio Storaro Paints with his Camera." *Moviemaker* (January 2008). http://www.moviemaker.com/articles-moviemaking/vittorio-storaro-cinematographer-caravaggio-20071127.

Gates, Chris. "Color Grading." *Videomaker* 28, no. 3 (September 2013): 53–56.

Kracauer, Siegfried. *Theory of Film: The Redemption of Physical Reality*. Princeton: Princeton University Press, 1997.

Langlois, Gerard. "Les montreurs d'ombres VIII.—Néstor Almendros: L'homme aux miroirs." *Lettres françaises*. September 23–29, 1970, 18-19.

Lessing, Gotthold. *Laocoön: An Essay on the Limits of Painting and Poetry*. Translated by Edward Allen McCormick. Baltimore: The Johns Hopkins University Press, 1984.

Maltin, Leonard. *The Art of the Cinematographer: A Survey and Interviews with Five Masters*. New York: Dover Publications, 1978.

Milner, Victor. "Painting with Light." In *The Cinematographic Annual*, edited by Hal Hall, 91–108. Los Angeles: American Society of Cinematographers, 1930.

The Moving Picture World. "An Artist in Camera Angles." April 23, 1927, 714.

Predal, René. "Les grands operateurs (III)." *Cinema 73* 172 (January 1973): 95–105.

———. "Les grands operateurs (IV)." *Cinema 73* 173 (February 1973): 87–95.

Revault d'Allonnes, Fabrice. *La lumière au cinéma*. Paris: Cahiers du cinéma, 1991.

Russell, Sharon. *Semiotics and Lighting: A Study of Six Modern French Cameramen*. Ann Arbor: UMI Research Press, 1981.

Storaro, Vittorio. *Writing with Light, Part One: The Light*. New York: Aperture, 2003.

Zone, Ray, ed. *Writer of Light: The Cinematography of Vittorio Storaro, ASC, AIC*. Hollywood: ASC Press, 2000.

Chapter 6

Hou Hsiao-hsien and Mark Lee Ping-bing

A Taiwanese Creative Team

Peter H. Rist

There have been many examples of long collaborations between great directors and excellent cinematographers in film history—for example John Ford with Joseph August on 14 films, intermittently from 1929 to 1948; Alfred Hitchcock with Robert Burks on 12 films from 1951 to 1964—and with some of these there has been recognition given to the importance of the cinematographer's collaboration—for example Billy Bitzer with D.W. Griffith for most of his films from 1908 to 1929; Eduard Tisse on all of Sergiei Eisenstein's films.[1] Although these last two directors' achievements have always been fully recognized, it is impossible to know the extent of these two cinematographers' stylistic contributions since their work with other directors was rare and scarcely known. Indeed with a few director/cinematographer collaborations it is recognized that a creative team was formed such that the cinematographer contributed importantly to the visual style of the director's films, most notably Gabriel Figueroa for Emilio "El Indio" Fernández on 22 films from 1943 (*Flor Silvestre*) to 1954 (*La rebelión de los colgados*) and Sven Nykvist for Ingmar Bergman on almost all of his films from 1960 (*The Virgin Spring*) to 1982 (*Fanny and Alexander*). Nykvist has contributed to a number of books including an introduction for a book sponsored by IMAGO, The Federation of European Cinematographers, in which he begins by stating that "The most important task of the cinematographer is to create an atmosphere," and goes on to quote Bergman saying in his book, *Laterna Magica/Magic Lantern* that "we are both restlessly captured by the problems of light."[2] As for Figueroa, there are almost certainly more books written on him than Fernández—seven that I am aware of—including most recently, a catalog for an exhibition organized for the Los Angeles County Museum of Art by the Mexican Televisa Foundation, *Under the Mexican Sky*.[3]

For my PhD thesis, completed in 1988, I argued that "John Ford" did not emerge suddenly as a genius film director, but developed his signature Hollywood Western and Americana compositional strategies over a number of years, perhaps learning from others who worked for him, especially cinematographers.[4] My research involved watching films shot by these cinematographers for other directors as well as Ford, and I focused a chapter on George Schneiderman's work for Ford at Fox between 1920 and 1934, and another comparing August's work with that of the one film Ford made at Goldwyn (*Arrowsmith*, 1931), a studio known for experiments in staging and filming in depth after 1928. By employing statistical analysis I was able to demonstrate that Schneiderman invariably filmed at eye level with few high- or low-angled camera shots, whereas August's tendency was to film off-eye level, and in high- or low-angle over 30 percent of the time, even though the subject matter of the films varied little from those shot by Schneiderman. At the time of writing my thesis, the use of statistics in order to more objectively analyze film style was uncommon and, although I was not arguing against the "greatness" of John Ford as a director, my suggestion was that he probably developed his style through working at different studios and with various, excellent craftsmen. In this chapter, I will argue that similar recognition could be given to the visual contributions that cinematographer Mark Lee Ping-Bing has made in shooting ten of the fifteen films directed by Hou Hsiao-hsien, from *A Time to Live, a Time to Die* (1985) until today. Moving in a somewhat reverse direction from my Ford/Schneiderman and August visual style research, I will recognize that Hou had already developed an individual style of long shot, long takes, where the invariably static camera often viewed his subjects at 90 degrees when Lee started to work for him, but that variations that occurred in this style, including a pronounced shift to the use of a slow moving and/or free-flowing tracking camera can be attributed to Hou finding an ideal cinematographic collaborator in Lee. Although my own visual impressions of the films will be central to my analysis, I will employ statistics as well as references to interviews with key players and to scholarly articles in order to present as objective an account as possible.

Hou was born into a Hakka family living in Guangdong province, mainland China in 1947. The family immigrated to Taiwan the following year, settling in the south of the island at Fengshan, where Hou then grew up. After compulsory military service, he entered the Taiwan National Academy of the Arts, where he studied film. A year after graduating, in 1973 he was hired by the Central Motion Picture Corporation (CMPC), where he effectively served his apprenticeship, beginning to work as an assistant director in 1974 and as a co-screenwriter in 1975. For a few years after getting his chance to direct in 1980 on *Cute Girl* (aka *Loveable You*) he continued to work as a screenwriter, an assistant director and an actor, most notably as the lead in Edward

Yang's *Taipei Story* (1985). His collaborative efforts, especially with Chen Kun-hou (his cinematographer on his first five features) and Liao Ching-sung (his editor on almost all of his films) were typical of the group of young emerging filmmakers stimulated by the production of two portmanteau films by CMPC, *In Our Time* (1982) and *The Sandwich Man* (1983) that initiated the Taiwanese New Wave. Even in his first three independently produced features, all starring Kenny Bee—the other two were *Cheerful Wind* (aka *Play While You Play*, 1981) and *Green, Green Grass of Home* (1982)—one can detect traces of Hou's later, mature style. David Bordwell notes that "in their relaxed simplicity" and restraint they are very different from Taiwanese commercial film productions of the 1970s, and he continues to state that, "Hou's willingness to take his players and camera outside and to let ordinary life crowd into his frame must have seemed refreshing."[5] In discussing low-budget production techniques during Hou's apprenticeship, James Udden notes how the director used longer takes, in part to allow "his actors more breathing room to perform."[6] One of the best features of *Green, Green Grass of Home* is the natural acting of the school children, and even at this early stage in Hou's career, one can appreciate his desire for "realism." Indeed in a recent interview, cinematographer Chen Huai-en, who began working with Hou as a script supervisor in 1983, confirms that, on *The Sandwich Man*, "Once he realized that actors could perform more realistically in a long shot, he stopped caring about closer shots."[7] In the same interview, Chen remarked that "I learned what Hou finds most important in an image from working with him for a number of years. The main thing is that he wants absolute realism and absolute spontaneity."[8]

Lee was born in Taiwan in 1954. His upbringing was similar to Hou's in that his father died young. Playing truant as a child, he would sneak into cinemas. (Hou was also movie-mad.) Like Hou, Lee also began working at CMPC as an apprentice along with his close friend Yang Wei-han, but he found very few cinematographers willing to reveal any of their secrets, except for Lin Hong Zhong (aka Lin Hung-Chung), whom he assisted on *Ku lian* (*Portrait of a Fanatic*, aka Unrequited Love) in 1982.[9] He then worked together with Yang on Wang Tung's *Ce ma ru lin* (*Run Away*, 1984), a film that Lee Daw-Ming interestingly calls a "realistic martial arts film."[10] The film was notable for its low light levels and it won the Best Cinematography Award at the Asia Pacific Film Festival in Japan. Wang Tung (aka Wang Toon) was already considered a major director in Taiwan, and Lee worked with him again as the sole cinematographer on *Yang chun lao ba* (*Papa's Spring*, aka Spring Daddy, 1985). In the same year, he worked for Hou as his cinematographer on the groundbreaking film, *Tong nien wang shi* (*A Time to Live, A Time to Die*).

Hou had first been recognized inside Taiwan when he was nominated for writing the adapted screenplay of Lee Hsing's *Zou an Tai bei* (*Good Morning,*

Taipei, 1979) by the 1980 Golden Horse Awards in Taipei and then the following year as Best Director of *Green, Green Grass of Home*. Internationally he jointly won the top prize, Golden Montgolfiere at the 1984 Nantes (France) Three Continents Festival for his first "new wave" feature, *Feng gui lai de ren* (*The Boys from Fengkuei*, aka, All the Youthful Days, 1983), and then again the following year for *Dong dong de jia qi* (*A Summer at Grandpa's*, 1984), but it was when *A Time to Live* won the FIPRESCI Prize in the Forum of New Cinema section of the 1986 Berlin International Film Festival, and later in the year screening at the New York Film Festival, that the director was really noticed, globally. One could also argue that it was with this film that Hou's signature style—static long takes, where the camera views the subjects in long-shot scale, often perpendicularly—became clearly noticeable. To be fair, it should be noted that, from the very first two portmanteau, New Wave films, location shooting tended to be accomplished in long takes, often using long lenses. With inexperienced crews, nonprofessional actors, low budgets and limited access to studio spaces, it was easier to shoot scenes in master shots without recourse to complicated multiple camera setups. One of the new directors, Yee Chin-hyen even suggested that the static camera predominated because the inexperienced camera crews "did not know how to execute steady travelling shots."[11] Although the static, long take, long-shot style, where subjects are often seen in full figure, not filling the frame, was initially a group style, Hou was the key practitioner. Udden has demonstrated how Hou's developing long take style was different from earlier practitioners such as Japan's Mizoguchi Kenji, Hungary's Jancsó Miklos and Greece's Theo Angelopoulos, in that there was relatively little camera movement in Hou's films. In eight comparative films he sampled from these three directors, more than 60 percent of the shots contain camera movement whereas the percentage in Hou's films actually declines from 45 percent for *The Boys from Fengkuei* to 23 percent for *A Time to Live*.[12] The female writer, Chu Tien-wen (or Zhu Tian-wen), is Hou's closest collaborator, being credited with the screenplays of all but one of the director's feature films since *The Boys from Fengkuei*, and she introduced him to Shen Congwen's autobiography during the making of this film, which influenced him greatly. Of this connection, Udden writes,

> Enamored of what he saw as Shen's distanced, detached perspective, Hou literally tried to capture that same feeling on film by constantly requesting that the camera be moved further and further back. Of the three hundred plus shots in this film, no less than one fifth of them have extreme long shot as the primary shot scale. Almost another fifth are roughly a long shot.[13]

According to Jeffrey Kinkley, who wrote his biography, Shen Congwen (1902–1988) was one of the greatest authors of modern Chinese literature, and for Emilie Yueh-yu Yeh and Darrell William Davis, his "native-soil fiction," "critical lyricism," and "imaginative nostalgia," in David Der-wei "Wang's terms struck a chord with Hou."[14] They continue,

> Through Zhu Tianwen, Hou found a new identity, which enabled him to rediscover a viable position and method—a poetics of autobiography—to rearticulate a worldview still in fashion. Shen's lyrical storytelling entails a spirit of evanescence, where nothing much happens. This view enables Hou to treat life, human relationships, and incidental encounters with subtlety and sensitivity.[15]

Barry Salt is the pioneer of using statistical analysis in film scholarship and he has self-published the book *Film Style and Technology: History and Analysis*, now in its second edition.[16] His work has also been at the core of the data included and regularly updated by contributors to the *Cinemetrics* website, founded by Latvian film scholar, Yuri Tsivian.[17] The *Cinemetrics* measurement of shot scale is based on Salt's system, illustrated at http://www.cinemetrics.lv/salt.php, where BCU (Big Close Up) frames the head only, CU (Close Up) frames the head and shoulders, MCU (Medium Close Up) is from the waist up (although the image shows from the stomach or chest-up), MS (Medium Shot), "includes from just below the hip to above the head of upright actors" MLS (Medium-Long Shot), "shows the body from the knees-upwards," and, although not defined by Salt, FS (Full Shot) shows the full body in the shot, LS (Long Shot), shows the body filling only half of the frame, while VLS (Very Long Shot) shows the actor small in the frame. For my own measurements of shot scale, I have tended to use the same scheme for closely framed shots (with ECU, instead of BCU), but I usually employ "MS" more widely (to include Salt's MS and MLS), while I reserve MLS for full body shots (FS). I use ELS ("Extreme") in place of VLS. Of course, many shots, especially those where a wide-angle (short focal length) lens is employed would display a range of shot scales, sometimes from CU to ELS, and in these cases, some averaging is necessary. In any event, most of my analyses here revolve around the measurement of average shot length (ASL) and the incidence of camera movement rather than shot scale, understanding that Hou has consistently resisted the use of CUs and MCUs throughout his career as a film director. Although we don't know to what extent long take shots were cut down in the editing room, nor how many alternate camera setups have been used by Hou and his cinematographers, we can certainly point to stylistic tendencies through objectively measuring and calculating data from the finished films. The ASL of films, or parts of films is simply calculated by dividing the running time by the number of shots.

The ASL of *A Time to Live*, is 24 seconds, already five seconds longer than the ASL of *The Boys from Fengkuei*. As mentioned above only 23 percent

(21 percent by my own calculations) of the film's shots contain camera movement, although many of these (mostly panning) movements are quite extensive or elaborate.[18] Before working with Lee, Hou was clearly developing staging in depth along the lines of what Bordwell calls "planimetric" composition, where the "background is resolutely perpendicular to the lens axis, and the figures stand fully frontal, in profile, or with their backs directly toward us."[19] This was probably new to Lee, and, although he doesn't discuss this aspect of working with Hou, there were times in *A Time to Live* when the camera was positioned close to the floor of the Japanese style family home/set, that could have been difficult to execute. This is one rare example of stylistic proximity to the films of the great Japanese director, Ozu Yasujiro. Another is a visual motif of telephone wires viewed in low-angle shots, marking transitions between scenes, that Noël Burch calls "pillow shots."[20] It is possible that Lee could have been partially responsible for these stylistic choices, since Hou claims that he had never seen an Ozu film until after he made *A Time to Live.*[21] Although Hou may well be close in spirit to Ozu, his visual style is radically different from the Japanese master's. For example, the ASL of Ozu's last six films (from 1958 to 1962) is just over seven seconds, and while there is only one example provided of shot scale analysis of these late Ozu films available through *Cinemetrics—Ohayu* (Good Morning, 1959)—it confirms the impression that MS and MLS compositions dominated his work, at 71 percent of all the shots.[22]

One of the ways Lee helped Hou on *A Time to Live* was his ability to work well with natural daylight and to incorporate household lighting when shooting interiors, while another was his understanding of color, which came from his childhood appreciation of Chinese paintings and porcelain. While working with him for the first time Lee surmised that Hou "had not been actively involved in framing and lighting questions before that."[23] He was impressed that Hou was always willing to improvise and not stick to a rigid script. In one outdoor scene that was supposed to be shot in sunshine, the beginnings of a typhoon was in evidence, but Hou directed Lee to film a 100-year-old tree swaying in the strong wind. By way of recognizing the director's desire to shoot regardless of the weather conditions, Lee says in the documentary, that for Hou, "every day's a fine day."

Hou's static, long shot, long take style became even more pronounced on his next film, *Lian lian feng feng* (*Dust in the Wind*, 1986), which, for Udden and myself, is one of the greatest achievements of the Taiwanese New Wave.[24] According to my calculations of the shot-by-shot breakdown I conducted on the 2001 Sino movie.com DVD, 59 percent of the shots in *A Time to Live* are "Long Shots" (LS and ELS), while, according to Udden, 104 of the 196 shots in *Dust in the Wind* (53 percent) either contain no people at all or are primarily "long shots" or "extreme long shots."[25] I suspect that Udden and I might have slightly different concepts of what constitutes a Long Shot. As I discuss earlier, for me it is where the film frame shows the full body of the human character(s), at least. In my analysis of the 2010 Central Pictures Corporation Blu-ray disc, I counted only 81 shots of *Dust in the Wind* that were mostly framed as medium shots (i.e., MS, MLS) or close shots (i.e., MCU, CU, ECU) and therefore not "long" shots, usually of one or both of the youthful, principal characters, Wan and Huen, whereas Udden would seem to have counted more than 90. Where a shot begins in Medium Close Up (MCU) and ends in Extreme Long Shot (ELS), or vice versa, it is difficult to determine what is the "primary" shot scale. In any event, in many shots of *Dust in the Wind*, the characters have their backs turned to the camera, hiding their facial expressions, and so, even if they are framed closely, we are very rarely seeing them as we traditionally do in close shots. In summary, Hou and Lee consistently framed the characters of both films at an unusually long distance. More significantly, "in terms of duration, the average shot length jumps to 33 seconds [for *Dust in the Wind*], a figure that exceeds most of [Jean] Renoir's films and matches many of Mizoguchi's famous works," while, differently from these directors' films, where camera movement is a key stylistic component, "less than one fifth of the shots (18.4 percent)" of *Dust in the Wind* contain camera movement.[26]

Without explaining exactly what he meant by it, Lee said that the "lighting arrangements that I developed on both films were intended to create visual contrasts while remaining realistic."[27] Apart from visual style, one of the remarkable aspects of both films is that their narratives tend to be episodic, hence "realistic," showing scenes in the lives of the young male protagonists drawn from autobiography —of Hou and screenwriter Wu Nien-jen, respectively—where the most dramatic events are downplayed or elided completely. Letters play a key role in *Dust in the Wind*, providing information, for example how Wan learns through a letter from his brother that his girlfriend, Huen has married a postman. Unlike "classical Hollywood" films where the larger-than-life central characters drive the narrative forward and where scenes are tightly connected in cause and effect patterns, these two films are structurally closer to those of the postwar Italian "neo-realist" movement where forces beyond the characters control determine their futures.[28] Motifs of Taiwanese landscapes, trains, timepieces (clocks and a watch), and meals with family and friends in *Dust in the Wind* allow for a retrospective understanding, linking characters and filling gaps in the narrative.[29] With respect to "realism," Udden argues that it was *Dust in the Wind* that marked Hou's "primal return to the very basic qualities of human life."[30] On the other hand, the distance of the camera from the human subjects and the slow, but fragmented pace, which seems at times more related to the natural world rather than the social world, provide a somewhat experimental edge to the film, both narratively and stylistically. Along these lines, Haden Guest argues that *Dust in the Wind* works as a reflexive, "meta-cinematic" text that brings "a heightened concern for *rhythm*, both as a structuring device and a larger thematic concept," as well as containing a "deliberate fracturing of time and space."[31] With the understanding that *Dust in the Wind* was both a relatively experimental work and arguably the pinnacle of the Taiwanese New Cinema, one could also understand that Lee's experience in working with Hou for only the second time must have been enlightening. Before working with Hou, Lee was fully competent technologically, and in an interview with Richard I. Suchenski he claimed that he "proposed even more daring visual ideas" to the director on *Dust in the Wind*, who "gave [him] room to explore" them.[32] As he says in the documentary film, it was during his work here that he learned to "integrate real light and color with people," and he credits the director with his understanding of human feeling and its reflection through cinematography.

The ending of *Dust in the Wind* is extremely interesting with regard to its emotional content, and is likely to have been a key scene in Lee's thinking about cinematography in a new more emotional way, where the "image can have a certain aura or flavor."[33] We see Wan lying on his army bed as his brother's letter is read in voice-over. Then a few brief silent shots at his home village show Wan and Huen's unhappy mothers together, followed by the newly married couple, and a medium shot of Wan still lying on his bunk, but now crying desperately; a show of emotion far exceeding anything that had preceded it. Strangely, this is followed by an extreme long shot at dusk of a line of trees at the bottom of the frame and a cloudy sky above. The camera tracks right as a guitar begins to accompany the movement. For those of us shocked by such an unfamiliar display of feeling, this acts to distance us emotionally and narratively from Wan, and a familiar, extreme long shot of the countryside, including a railway signal, followed by an empty interior long shot of the doorway and entrance corridor of the family home adds to the pause effect. Wan is wearing a brightly decorated shirt that Huen, a seamstress had altered to fit him, and as he turns the corner to look into a room, a cut shows his mother asleep on the floor. The final exterior scene in the family garden, with forested hills in the background, is composed of two shots, where Wan's grandfather talks of the difficulties in growing potatoes and ginseng this year. The film's final shot is a long take of the surrounding hills and the sea beyond, accompanied by the quiet theme music that has only occasionally been heard. For some of us, this gentle and hopeful reconciliation between grandfather and grandson, in the natural environment of Taiwan, is incredibly moving, all the more so because of the filmmakers' understanding that we don't need close ups.

Lee spent most of the next eleven years working as a director of photography in Hong Kong. He worked on a lot of action films where he learned to work fast and shoot with a handheld camera, including *Fong Sai-Yuk 2* (The Legend II, 1993) directed by Cory Yuen with whom he won his first Golden Horse award for Best Cinematography. He was paid by the hour. But, he also worked with more serious filmmakers especially Ann Hui, for whom he shot three features, including, *Nu ren si shi* (Summer Snow, 1995), which brought him his second Golden Horse award. At the end of this period, he shot *Tempting Heart* (1999) for the great Taiwan-born actress Sylvia Chang, who claimed in the documentary film that, apart from learning speed and precision, Lee was frustrated by his Hong Kong experiences. He returned to Taiwan to work with Hou on the second of three key films dealing with Taiwanese history, *Xi meng ren sheng* (The Puppetmaster, 1993), which, for some of us is one of the greatest films ever made.[34] Indeed, this film marks the peak of Hou's static, long take style, with an ASL of 82 seconds, twice as long as any of his previous films and where only 14 percent of the 100 shots have "noticeable" camera movements.[35] Two of the most important contributions made by Lee to this film are in the realms of color and interior lighting. In exploring history through the everyday life of Li Tianlu, the "puppetmaster" of the English title, and in showing many puppet performances in tableau style long takes, Lee recognized that the film required richer colors, and according to Lee, Hou gave him the responsibility "for determining the overall color palette for the film."[36] In the same interview, Lee claims that he had to work with "dimmer" lighting than ever before, and yet still give the impression that the lighting sources were natural (daylight) or from household fixtures of the period (1911–1945). Udden argues that Lee's lighting arrangements seem to be much more "directional" than in previous films, where "in some cases" he even used "backlighting that completely silhouettes the main figures on the screen," and that whereas the dimness "challenges the viewer in some respects, these interior views are hauntingly and exquisitely beautiful on the big screen." [37]

The third of Hou's films dealing with Taiwan's history, *Hao nan hao nu* (Good Men, Good Women, 1995) was shot by Chen Huai-en, and, while continuing to extend Hou's ASL (to 109 secs), it contained much more camera movement, especially in the contemporary sequences that compare modern-day youth unfavorably with that of the activists left during the country's "White Terror" when some 900 Communists were executed between 1949 and 1960. I have conducted a shot-by-shot breakdown of the Sinomovie.com DVD (no date) and I found that only 12 of the film's 56 shots contained no camera movement at all (21 percent), while another four shots had very slight reframing movements. Almost all of the scenes are represented by single long takes. In fact, most of the camera movements in the film, usually panning and tilting, with some slow dollies are readjustments to follow the movements of the one or two main characters in the shot. Some of the scenes in the contemporary story are filled with movement while most of the scenes in the historical episodes have camera movement just at the beginning or end of the shot. Clearly, Hou was changing his style here, but the employment of camera movement is rarely of interest aesthetically, and simply follows the action. I suspect that Hou may have been displeased with Chen's work, and he has stated that *Good Men, Good Women* is his least favorite film, perhaps because of the cinematography.[38]

But it is with Hou's next film, *Nan guo zai jian, nan guo* (Goodbye South, Goodbye, 1996) that real stylistic changes begin to occur, no doubt spurred-on by Lee's capabilities. Set completely in contemporaneous Taiwan, much of it is shot on the road, in motion. By my calculations of a shot-by-shot breakdown I conducted of the Sinomovie.com DVD (no date), only 10 shots (approx. 15 percent) didn't contain any camera movement other than very slight reframing, although I count "traveling" shots where the camera is mounted on the moving vehicle, while the world goes by outside (the car or train) as "moving camera" shots, even though, technically, the camera is not actually moving along the ground. There are also a number of long take, deep tracking shots, including one with Lee's camera presumably mounted on the back of a car, to show two motorcycles ridden by the protagonists, Kao (Jack Kao) and "Flattie" (Giong Lim) on country roads in the south of Taiwan near Chiayi City. Most noticeably, though, Lee was called upon to adopt his Hong Kong-learned handheld camera skills to shoot a Hou-directed film, and there are many shots where the camera pans and tracks to follow the action, which although more conventional, stylistically, also seem to be improvised. Indeed this is the first of Hou's films where, to quote the title of Lee's documentary, there are "fleeting moments." At the beginning of the film, Kao, Flattie and his girlfriend, "Pretzel" (Annie Shizuka Inoh) are travelling on a mini-train (the sound of which had played over the credits). After their arrival at Pingxi, the action shifts to the contiguous space of a bar/restaurant with a billiard table (a typical Hou location), where the offscreen sound of trains is continually heard. After getting into a fight where Lee's camera is called upon to move suddenly, "Flattie" appears on the roof. Down below are the train tracks, and, for no apparent reason the camera pans away from him and tilts down to show two women, who otherwise don't appear in the film, boarding a mini-train. It's as if the film suddenly becomes an observational documentary, and, even more strangely, the next shot is filmed from the front of the train going into a tunnel. Hou (sometimes prompted by Lee) made many other strange stylistic decisions in *Goodbye* including the choice of green and deep pink tinted shots, which are improbably justified as being Point of Views (p.o.vs) through a motorcycle helmet and sunglasses, respectively.

Beginning with *Hai shang hua* (Flowers of Shanghai, 1998), Lee has been the cinematographer on all of Hou's subsequent features. This also marks the point where both director and cinematographer become truly international with their projects, and where their joint style becomes modified with almost every project in ways that match the changes in subject matter. For the first time, Hou moved beyond Taiwan, and into the distant past: late nineteenth-century Qing Dynasty Shanghai and the "flower houses" of prostitution, frequented by idle rich men. They had gone to China to shoot some of *Puppetmaster*, but for this film, since they had to try and recreate interior architecture, decor and costumes that no longer existed, and as the Chinese authorities objected to the decadent subject matter, they built sets in Taiwan. Chu T'ien-wen's screenplay was based on Eileen Chang's Mandarin translation of Han Bangqing's densely detailed two-volume novel, so that the filmmakers could visually interpret varied aspects of life in the "flower houses." Following his usual desire for realism, Hou wanted oil lamps from the period to light the sets, but Lee knew that this wouldn't be possible. Nevertheless, he brilliantly produced lighting setups that suggested what he termed to be a "glamorous realism," evoking the "greasy light and silken textures that *Flowers of Shanghai* evoked in [Hou's] mind."[39] Every scene ends with a slow fade to black, where the oil lamp flames are always the last images to fade. Indeed, the belief that the lighting emanates only from the oil lamps is so strong that one critic argues that the "flames" of these lamps become "concrete cinematic subjects that slowly drift into and out of view."[40] In an interview conducted for this book chapter through e-mail, Lee talked about how difficult it was to give this impression, especially working in small spaces. He stated that:

> In order to provide the maximum space for actors to perform and maintain the most realistic balance of light and shadow, the lighting designer and I designed a spotlight with a 2K tungsten lamp, wrapping it with poly on all sides. We put these spotlights above the oil lamps, high enough to stay outside of the frame. The light from a spot was concentrated to the center of an oil lamp on the table. We wanted to enhance the brightness of the reflected light from the oil lamps, enough to highlight details of the actors and the scene. The poly wrapped around the spotlight was also affected by the brightness and produced some faint light. I used this diffuse light for some dim details in the environment, and I added some colors to the scene with color gels.[41]

It is important here to recognize that this film was surely one where its overall sense of beauty was created through the very close partnership of Hou and Lee. Going further, Udden suggested that *Flowers* "could be the crowning achievement of Mark Lee's illustrious career" as one of the "most beautifully lit films ever made anywhere."[42]

Every shot but one of *Flowers* contains camera movement—slow pans and tilts, with imperceptible deep tracking movement to get closer to the characters—and the ASL is close to 160 seconds, the highest ever (still) in Hou's oeuvre, but the majority of the film is rendered in Medium Shot scale, unusually for the director. With low light levels, elaborate costuming and with all the men sporting shaved foreheads and queues, it might have been necessary for the camera to be closer to the many male subjects for audiences to clearly identify their faces, and Hou's deliberate strategy here also reflects how it was only in spaces such as the "flower houses" that men and women could interact so intimately. In an interview with Li Dayi, the director noted how he discovered in *Flowers* that the camera didn't necessarily need to be far away in order to achieve a sense of distance if everything is quiet on the set and where there is a trusting fondness between cast and crew. In Udden's translation of this discussion he wrote that the "camera becomes just like a person next to the action. Sometimes this person looks one way at a group of people or a person, then will shift to the other side when he or she hears something there."[43]

The opening shot of *Flowers of Shanghai* lasts for more than eight minutes, and is probably the longest take in any of Hou's films. It also provides an introduction to the style of the film as a whole, with the importance of the lighting, the complex staging of characters' dialogue and their movements along with the slowly arcing camera, panning and tilting to shift the viewer's attention from one character to another. Interestingly, in a recent (2013) interview Lee has stated that

> at an early stage [of the film's production], Hou asked if I would like to try filming them with a moving camera. I was, of course, very excited, so I started panning the camera slowly but aimlessly. Breaking with convention, I decided to follow those who are listening rather than those who are talking, trying to capture their facial expressions. . . . Hou left all decisions about what to capture to me and he asked me to move the camera based on what I felt was the rhythm."[44]

It is here that I believe we can understand the essence of the collaboration between director and cinematographer, with Lee being able to improvise after Hou, working closely with screenwriter Chu, has planned the characters' dialogue and their interactions. In Udden's excellent analysis of *Flowers of Shanghai*, he writes at length on the opening banquet shot (almost three book pages), discussing the "density of detail," with "three distinct planes [of action] competing for the viewers' attention," and he concludes:

> This opening shot, arguably the acme of Hou's indirect narration, has to rank as one of the most densely and yet brilliantly wrought in the annals of film history; this long take is almost a film—even a world—onto itself.[45]

From *Good Men, Good Women* on, all Hou's films are international coproductions, a strategy necessitated to financially support films that might only show in film festivals and a few "art houses" in Europe and North

America. Indeed 2000 is the year that Lee began his own international travels with the Vietnamese film, *Mùa hè chieu thang dung* (The Vertical Ray of the Sun), directed by Tranh Anh Hung, and, most notably, Wong Kar-Wai's *Fa yeung nin wa* (In the Mood for Love), a Hong Kong production (coproduced by Paradis Films, France), the making of which he joined in Bangkok, when the original cinematographer, Chris Doyle had to leave. Apparently, Lee shot approximately two thirds of *In the Mood for Love*, and, although it is impossible to determine exactly which shots each cinematographer was responsible for, it is clear that the "glamorous realism" associated with Lee's work with Hou is on display perhaps more here than any other of Wong's films.[46] Wong, of course is known for directing glamorous commercials as well as films, but the silken beauty of female star Maggie Cheung's *cheongsam* dresses, subtly lit while she walks along the sidewalk in slow motion, laterally tracked by the camera, is evocative of Lee's work on *Flowers of Shanghai*. Lee loved to use tracks, whereas Doyle was more associated with dynamic handheld camera shots, varying running speeds, and step printing.[47] Indeed Cheung's character, (Mrs) Chan Li-zhen is originally Shanghainese, and the intimacy of the camera—she and her co-star, Tony Leung Chiu-wai are often framed in Medium or Medium Close-Up—along with the absence of any depictions of sexual activity, are also reminiscent of *Flowers*. Nevertheless, there is an erotic charge to their scenes both apart and together, which is quintessentially Wong's. With critics, at least, *In the Mood for Love* is arguably Wong's most popular film. IMDb lists 43 award wins and 40 nominations, and it was also Lee's most successful project in terms of prizes: he shared at least six major cinematography awards with Doyle, including his third Golden Horse Award and the U.S., New York and National Society of Film Critics' awards, as well as sharing the Cannes Technical Grand Prix with Doyle and William Chang.

Lee's very next project was also Hou's next film, *Qian xi man po* (Millenium Mambo, 2001). Lee maintained Hou's long take style (ASL 97 secs), while continuing to move the camera a great deal (in 82 percent of the shots).[48] As with *Flowers of Shanghai* the camera often frames the characters in Medium Shot, but, yet again, the visual style has been modified slightly in accordance with the subject matter. At the beginning of his career, Hou relied on the long lens to film his actors, unnoticed by passersby on the streets, and then, as David Bordwell discusses, he employed the telephoto lens in his new wave films, especially to film dinner table scenes, with "his players" arranged "in horizontal layers."[49] However, there is a great deal of depth in some of the shots in *Goodbye South, Goodbye* and *Flowers of Shanghai*, where Lee must have been using wider angle, shorter focal length lenses, while the lack of depth in *Millenium Mambo*, especially in the scenes staged in Vicky's (Shu Qi) apartment, where shallow backgrounds and foreground objects, including alcoholic beverages are rendered deliberately out-of-focus, is noticeably different. Much of the time, the framing is planimetric, which together with the use of a long lens and/or a wider aperture tends to flatten the perspective while the extremely cluttered arrangement of primary colored objects increases the confused, sometimes unpleasant visual experience for the viewer that Bordwell compares to the films of Joseph von Sternberg, suggesting that this new style "conjures up a hazy, drug-fueled hedonism."[50] In Chang Chuti's interview Lee stated that he normally prefers lenses with wider apertures and he also stated his general preference for using tracks over a Steadicam:

> Images shot with dolly tracks and Steadicam express very different feelings. Using dolly tracks, I maintain a certain distance from actors, giving them more space to perform freely and not disturbing their acting. I can also do a long take, and the change of frames, images and rhythm can be easily adjusted. I am more in control since I operate [the dolly] myself. By contrast, Steadicam cannot easily achieve these things, and I also need to spend more time to communicate [with the film's crew] and increase the budgets.
>
> When it is appropriate, I will use the Steadicam too. In the opening sequence of *Millenium Mambo*, Miss Shu Qi walks along a roofed footbridge. I used Steadicam to film this long take, but I still operated it myself so that the images deliver another feeling.[51]

One could argue that Hou's representation of restless, contemporary youth, beginning with *Good Men, Good Women*, is unfairly negative (especially toward young women) in contrast to his humanitarian depictions of people living in the past, but, it is clear, that with *Millenium Mambo*, Lee and Hou modified their visual style appropriately in accordance with the subject matter, including the intensification of bright, snappy colors, moody and neon lighting, and more dynamic movement.

In 2001, Lee was hired by the visually sophisticated 5th Generation, mainland Chinese director, Tian Zhuangzhuang to film a remake of the Chinese classic, *Xiao cheng zhi chun* (Spring in a Small Town, 1948, dir. Fei Mu). With exactly the same title, Tian's remake is known by the English title, *Springtime in a Small Town* (2002). Lee laid tracks for both interior and exterior scenes and, by my count, only three of the film's slightly more than 90 shots had no camera movement at all. None of Tian's films made before or since have contained anything like this amount of camera movement. Significantly, two of these three camera movement-less shots in *Springtime* are the film's last two shots. They are completely static, when after suggestions of change to the decaying, once gloriously rich provincial household, life goes back to its norm of stasis.[52] For the purposes of telling his story objectively, Tian wanted Lee to keep the camera at a distance of about eight meters from the characters, perhaps influenced by Hou/Lee's "realist" approach. And for the film's longest six minutes and twenty-five seconds. shot, Lee built a semicircular track around a dinner table, to film the four principal characters playing drinking games, lit by oil lamps—the electrical power was out in the story—with the camera in almost continuous, sometimes imperceptible motion: a nearly identical, though less complex reprise of Lee's work on *Flowers of Shanghai*.[53] Indeed, throughout *Springtime* one is struck by the brilliance of the cinematography performed in low daylight and artificial lighting conditions.

After shooting his first all-PRC production, Lee moved to Europe for the first time to work on Gilles Bourdos' *Inquiétudes* (A Sight for Sore Eyes, 2003), a French film. He and Hou then got together again to make *Kôhî jikô* (Café Lumière, 2003) in Japan, as a centenary tribute to the great, long departed Japanese film director, Ozu Yasujirô. The Taiwanese context is maintained by having the central, female Japanese character, Yôko (Hitoto Yo) pregnant by a Taiwanese man, and looking for a Tokyo café that was frequented by a (real) Taiwanese composer, Jiang Wen-ye, who had died in 1983. Shots following her walking are reminiscent of the long lens location work of Hou's first two films, and Yoko's travels on the Tokyo subway and commuter trains allow for Lee's free-flowing camerawork to reveal the enormity and complexity of these railway networks. In one remarkable extreme-long shot camera set-up (seen two or three times) we see three different train lines, aligned vertically, and a road bridge above them all in the far background. Sensationally, in one of these shots, three trains are viewed, simultaneously passing one another.[54] Although the ASL is down from *Millenium Mambo* to 66 seconds and although there is a slightly lower percentage of shots with camera movement (76 percent) than in Hou's previous film, the movement-filled nature of all of the train scenes injects a real dynamism to *Café Lumière* that is unlike any other Hou film.[55] It is as if his own love of trains that is suggested in earlier films, especially *Dust in the Wind* and *Goodbye South, Goodbye*, is allowed to come to the fore. One could also argue that while the exhilarating train scenes invoke the extreme modernity of urban Tokyo, that they could also suggest the entrapment and alienation of the youthful protagonists (not a new theme for Hou). Perhaps, Hou could have been expected to modify his style to suit a homage to Ozu, with low camera positions and more intimate framing of characters in interiors, but rather than a stylistic homage, the allusiveness can be found in the narrative, through the relationship between a father and a daughter (Yôko) that had dominated Ozu's last two films, *Kohayagawa-ke no aki* (The End of Summer, 1961), *Sanma no aji* (An Autumn Afternoon, 1962).

After *Café Lumière*, Lee worked again in the PRC for a first time, female director, Xu Jinglei, on *Yi ge mo sheng nu ren de lai xin* (Letter from an Unknown Woman, 2004). This film was based on the Stefan Zweig novel, *Brief einer Unbekannten* that had been adapted into the famous Hollywood film, directed by Max Ophuls, *Letter from an Unknown Woman* (1948), and for this Lee won his first official, China Film Association, Golden Rooster Best Cinematography award. Then, Lee again worked with Hou on the anthology film, *Zui hao de shi guang* (Three Times, 2004), literally translated as "Best of Times." It is set in three different time periods of Taiwan's history and effectively looks back over Hou's filmmaking career, both stylistically and thematically. The first part, "A Time for Love" is set in 1966 Kaohsiung, the second largest city, in southwest Taiwan, close to where Hou grew up. The second part is ironically entitled "A Time for Freedom," and is set in 1911 Dadaocheng, the historic port region of Taipei, while the final part, "A Time for Youth" is set in contemporary Taipei. In each part, the central female and male characters are at some stage of a love relationship, and are played by Shu Qi and Chang Chen, respectively. "A Time for Love" is clearly somewhat autobiographical, and is related visually to Hou's first "new wave" films, with deep focus long takes inside a pool room, where May (Shu Qi) works as a hostess, and is asked by Chen to play snooker with him just before he enlists in the army. It is, by far, the most romantic of the three stories. In some ways, "A Time for Freedom," is stylistically unique, as Hou attempted to evoke the silent film epoch, in the early twentieth century, with no dialogue, and by inserting title cards. Mr. Chang, the son of a tea plantation owner helps to buy out the contract of a young "flower girl" (prostitute) but his own favorite courtesan (Shu Qi) is not able to become a legitimate concubine, when he chooses politics over the personal and joins a Chinese revolutionary publisher. In both subject and style, this story is reminiscent of *Flowers of Shanghai*, with similarly ravishingly beautiful interior decor and costumes, and where oil lamps are apparent light sources, but underscored here by a sense of entrapment with background walls, opaque glass windows and room dividers limiting spatial depth, and with Lee's intermittently panning camera not allowing for a feeling of openness, ultimately delimiting the unnamed woman's possibilities of "freedom." The modern story is closely related stylistically to *Millenium Mambo*, but with an even more dynamic and neon-lit expressive account of urban youth. However it lacks the previous film's optimism. As the press kit relates "A Time for Youth": "Epileptic and losing sight in her right eye, Jing [Shu Qi] is a singer, lives with her mother and grandmother and also has a woman lover: Micky. Zhen [Chang Chen] works in a digital photo shop and lives with his girlfriend Blue. When Blue finds out that Zhen has fallen for Jing, she hits the roof. . . . Where can the four of them go from here? Maybe none of them will get any peace this side of the grave."

Not only was *Le voyage du ballon rouge* (Flight of the Red Balloon, 2007) Hou's next film, the first to be shot outside of East Asia, it was also made as a tribute to Albert Lamorisse's French short film, *Le ballon rouge* (The Red Balloon, 1956) and in being produced by the Musée D'Orsay, built at the turn of the twentieth centuries and housing the largest collection of impressionist and post-impressionist paintings in the world, the film, unsurprisingly pays tribute to Paris, France, and, in my view, French impressionism. Cast against type, Juliette Binoche plays a chaotic character, Suzanne who hires a Taiwanese film student (Song Fan) to look after her young son, Simon while she works as a puppeteer. Although the film is made in France, one can see many connections with Hou's roots, including in the film's style. The ASL is still very high, 75 secs, and much of the film is shot in long-shot scale.[56] It seems to me that, of all of the films that were shot by Lee, this is the one where his free-flowing approach is most evident. It is reflected in the fact that there is camera movement in 98 percent of *Flight*'s shots and that Lee had to follow the occasionally uncontrolled flight of a balloon. What I term "chaos" can mostly be observed in the state of Suzanne's life in her apartment, but also, I argue it is in this film's camera movement. Additionally, a sense of "freedom" is surely also found in impressionist painting, and so-called French "impressionist" films of the 1920s. Although, one would never claim *Flight* or any other of Hou's films to be "subjective" in using "optical effects such as superimpositions [to] imply characters' thoughts and moods," one finds style elements that have rarely appeared in Hou's previous work, including low and high angled shots, dizzying movement through train windows, dissolves and momentary superimpositions.[57]

It makes sense that the camera will look up to the sky to follow the progress of a balloon, and in *Flight*'s first seven shots, lasting seven minutes and twenty seconds, the camera moves most of the time. The film fades in on Simon holding onto a Metro station fence as he looks up to the sky and talks to himself about a balloon. We are on location with passersby not apparently aware that they are involved in a fiction film shoot. Lee's camera tilts down and pans right as Simon makes his way down the Metro's steps and then tilts up and pans right to a tree with some of its leaves slightly out-of-focus and shining yellow in the sunlight, with a red balloon behind it. The camera then follows the balloon down close to the station entrance and back up to the sky and around the tree. In the second shot we are in a dark tunnel with the camera mounted in the driver's cab of a train as it emerges. The third shot seems to begin as the first left off, but as the camera tilts down to continue the movement of the balloon we realize that we are in a different location and as the balloon disappears into a building a train arrives from the right and follows it into the station. In the fourth shot, like the first, a very long take, the balloon is now much closer to the camera and hanging over a station platform, and with a train sitting at the opposite platform, another train arrives and the wind it creates blows the balloon even closer to the camera. At least one passenger touches the balloon presumably to help Lee keep it in frame. The fifth shot shows Simon inside a train car standing close to a window looking out at the balloon, and the train then moves and the cityscape flashes by through the window. The sixth shot again follows the balloon in the far distance moving right along a railway bridge, while the seventh shot is filmed from eye level on a busy street showing buses stopped and passing, eventually revealing Song standing, turning and walking a bit on the far sidewalk as if she is perhaps lost. In the second, fifth, and seventh shots there is a great deal of attention focused on reflections in train, bus and car windows, even on the balloon in the fourth shot. I suggest that this visual play on reflections, and of sunlight on the leaves is indicative of Lee and Hou using cinematography to express themselves, being in Paris, the "city of light," where impressionist painters, whose work had also been grounded in reality, would play with light and color (in particular) to express their emotions.

Much of the midsection of *Flight* takes place in Suzanne's apartment or in that of her downstairs neighbor, where Simon is learning to play the piano. In one remarkably long take (four minutes and fifteen seconds), where the camera is mostly viewing Suzanne's living room and kitchen behind it at a 90 degree angle, the shot begins on the reflection in a slanted mirror or perhaps a sheet of glass fronting a painting (it is not clear) of a table lamp and Song sitting at a desk. The camera slowly pans left to show the scene no longer in reflection. The camera placing, its slow tracking movements and the modern setting are reminiscent of apartment scenes in *Millenium Mambo*, except that here, the natural light from a doorway on the left of frame and through the kitchen window in the background from the right create a much more positive image of urban living. Most remarkably, Lee managed to balance the two bright sources of natural light with the warm glow of artificial light, apparently emanating from the central table lamp set against the back, living room wall. As with *Flowers of Shanghai*, the lamp itself attracts the viewer's attention, almost as if it is a character.

Near the end of the film, the red balloon reappears, and the film climaxes with a series of high and low-angle shots, and striking reflections. We see the balloon flowing over Paris rooftops and then Simon climbs up the steps to his attic bedroom in high angle. We see the balloon through his skylight in low angle. Song is reflected in Suzanne's kitchen window as the balloon passes by, and then, at a later point in time, perhaps the next day, we view a group of school children in very high angle long shot taken through the misty roofing glass of the Musée D'Orsay. Now inside the museum, Lee rack focused from the reflections of the school group in the glass of a painting, again in high angle, to the painting itself, Félix Valloton's *Le ballon* (1899). The painting shows a child chasing after a red ball on sunlit sandy ground, viewed in high angle. Along with the light and the blocks of color, "high angle" becomes a discussion topic for the group. Simon is singled out as the subject of the camera's gaze, and he is distracted by the film's red balloon, viewed again in very low angle through the roof of the museum. The film ends with an extreme long shot of the familiar Paris skyline, including the Notre Dame cathedral, in hazy, but bright sunlight. Along with *Flowers of Shanghai*, *Flight of the Red Balloon* is arguably the most visually complex example of a film where Lee's cinematographic style perfectly matches Hou's narrative form and subject matter.

It is not unusual for Lee to take on three or four projects every year, and after *Flight* in 2007, he shot Jiang Wen's third film as a director, *Tai yang zhao chang sheng qi* (The Sun Also Rises), about which the actor/director marveled at the way Lee wanted to shoot during a snowstorm in the desert even though he was unaccompanied by any other crew members, providing another example of his preference to be a lone cinematographer. He worked again for Tranh Anh Hung on the filmic adaptation of Murakami Haruki's novel, *Noruwei no mori* (Norwegian Wood, 2010), for which he won his first Asian Film Award as Best Cinematographer. Most interestingly, for me, Lee shot two more films for Bourdos in France, the second of which, a biopic on the great impressionist painter Pierre-Auguste Renoir, gained him nominations for the French film industry César and the American Society of Cinematographers' Spotlight award (both in 2014).[58] I clearly remember watching *Renoir* on its general release, and realizing immediately that, although we were watching a digital copy, it must have been shot on 35mm film. Like *Flight*, this film was even more obviously made as a visual homage to impressionist painting, with the delicate play of light shining through the trees at the Rayol-Canadel-sur-Mer locations in the south of France, and where the sense of (film) grain rather than (digital) pixels glimmers.

It is hard to imagine a stronger visual collaboration between a director and cinematographer than Hou and Lee's *Nie yin niang* (The Assassin, France/Taiwan/Japan, 2015). Based on a Tang Dynasty short story, which is by necessity simple in narrative terms, *The Assassin* is stylistically very complex. A *xia nü*, woman warrior, Nie Yinniang (Shu Xi), working for the Tang Court, is sent by her master, Princess Jiaxin (Sheu Fang-yi), a Daoist nun, to kill a cruel and corrupt governor. On seeing him playing with a young boy (probably his grandson), she refuses to kill him, and is punished by being sent back to her family home in the Weibo region to kill her own cousin, Tian Ji'an (Chang Chen), who, as the head of the Weibo army, is scheming to attack the emperor. Yinniang and Ji'an were once betrothed, and in her childhood she loved his mother, her aunt, Princess Jiacheng (the nun's twin sister, also played by Sheu) who taught her music and a love of nature. On her return home, Yinniang remembers, in flashback, a scene where her aunt plays the zither. Initially, Hou wanted to make all of *The Assassin* on 16mm film, using only Bolex spring-wound cameras. Both Hou and Lee were interested in the restrictions implicit in the use of such cameras. For example, although one could shoot at a variety of transport speeds, from 8 to 32 frames per second (fps), at the normal running speed for sound film (24 fps), long takes would be restricted to only 17 feet or 28 secs allowed by the spring mechanism. Although Hou likes to work within "limits," in the end they decided that the challenges were too great and they would shoot everything in 35mm.[59] Nevertheless, it is possible that the four shots of the flashback were filmed in 16mm (and possibly the preceding shot of a group of travellers on horses, that seemingly includes Princess Jiacheng). The image is extremely grainy, and for the three close shots of the princess and the final shot of the four white peonies, the aspect ratio suddenly shifts from the old academy ratio of 1.37:1 to the new widescreen standard ratio of 1.85:1. The stylistic decision to choose a grainy appearance and a wider frame draws our attention to these shots that are marked as the female protagonist's subjective memories, and perhaps resonate more strongly than any other shots in the film, with emotion.

It follows that in classic *wu xia* style, going back to King Hu's *Da zui xia* (Come Drink with Me, Hong Kong, Shaw Bros., 1966), the action scenes of *The Assassin* are cut very fast. Nevertheless the overall ASL is still 29 seconds, showing that the majority of the film is presented in Hou's favored long takes. *The Assassin* begins with a black-and-white prelude of Yinniang, initially killing a horse-riding general with her dagger and then being instructed by Princess Jiaxin at her cliff-side temple to assassinate the governor. Apparently, none of the film was shot on black-and-white stock so, the color would have been removed digitally.[60] However, Hou claims that no other color correction or visual adjustments were made digitally, as he said to Aliza Ma that, "With *The Assassin*, all the 'effects' were applied directly on the camera, and not in the post-production process."[61] Perhaps the most remarkable scene of Lee's film camera pyrotechnics is where Yinniang watches Tian while he is visiting his favored concubine, Huji (Hsieh Hsin-ying) in her quarters. The sense of intimacy is heightened by having series of drapes, some of which are translucent and reflect the lights of candles, being moved (by unseen hands) in front of and away from the lens. Lee had begun to use plastic sheets to reflect light in 2003, filming *Inquiétudes* in France. Here with *The Assassin*, he also gradually changes the focus, while the camera is always on the move, imperceptibly tracking back and forth, while Yinniang is occasionally glimpsed in the background also hidden by another set of drapes. The slow moving, lateral tracking camera aesthetic that Lee introduced to Hou's oevre with *Flowers of Shanghai* and developed while working with Tian on *Springtime in a Small Town* reaches spectacular and beautiful fruition here. Although Lee has stated in an interview conducted for the major retrospective series of his work at New York City's Museum of Modern Art that "I think a film belongs to the director," and how he tries "as much as I can to understand what each director wants to pursue," he continued to state,

> I then try all I can to realize that vision. For example, I know that if I make a powerful and moving long take, Hou Hsiao-Hsien will use it. So when I collaborate with him, I work very hard to design these takes. On the other hand, when I collaborate with Ann Hui, I don't use this same approach. Hou Hsiao-Hsien is a director who likes to arrive on set, observe the environment, and then decide how to film. Ann Hui is a director who likes to plan ahead the actor's movements and how the camera should compose. So although they are very different, I still try as much as I can to capture light and shadow realistically in my own way."[62]

Another example of impressive visual style in *The Assassin* is Hou and Lee's treatment of landscape. The film has two conclusions, both occurring outdoors. The *wu xia* ending finds Princess Jiaxin, dressed as usual in white, sparring with her pupil prior to their departure, while holding a short whip/ fan rather than a sword or dagger. We had become familiar with Yinniang using her martial arts skills for defense rather than attack, and although there is a glimmer of suspense in the scene, and it is cut very fast like other fighting scenes in the film, we understand that there is no way the nun would harm her pupil no matter how much she has been disappointed by Yinniang's reluctance to follow orders. It is the last fight, and it is relatively playful. Just before this, the two had met on the top of a hill in extreme long shot, with Jiaxin looking left over a valley, and with majestic mountains behind, in layered stages of depth. One can imagine that the director/ cinematographer team could have waited a long time for the mist, first to arrive to obscure the valley and part of the background, just like a classical Chinese landscape painting, and then move to reveal and obscure other parts of the view. In the foreground there is a strange looking, camel hump-shaped hill covered in orange/red, autumnal vegetation. In an earlier scene featuring the quarters of the white-bearded wizard, Kongkong'er (Jacques Picoux), working for Tian's wife, Yuanshi, puts a spell on concubine Huji. In the background one can see a creatively imagined Tang Dynasty landscape painting, which contains aspects of the fantastic, but looks for all the world to have been based on the actual site where Yinniang and Jiaxin meet, above. Interestingly, in an interview with Suchenski, Hou noted that he didn't appreciate Chinese landscape paintings when he was young because they were "too far from me."

> Now I am able to appreciate them because I realize they are based in realism. This became apparent after I started shooting films in China and discovered that these paintings depict landscapes that I actually experienced in person. The mountains, the highlands, the fogs, and the trees, for example, are very realistic, not impressionistic.[63]

Impressionistic smoke and deception are associated with Kongkong'er, whereas the natural quality of mist is associated with Yinniang. So perhaps

Hou chose to show a stylized painting in the dark home of the wizard in contrast to a miraculously beautiful and bright, real landscape occupied by Yinniang and her master, while recognizing the former is based on the latter.

The second conclusion of *The Assassin* involves Yinniang accompanying her new friend the "Mirror Polisher" (Satoshi Tsumabuki) from a farming village through a pastoral foreground with another spectacular, mountainous landscape in the distance, on their way to Xinru, which Li Cheuk-to told me could be an old Chinese name for Korea (although I can find no evidence of this). It is implied that Yinninag has given up her role as a woman warrior, and is at peace with the natural world. The film's last shot continues until the small group walking with their horses, disappear into the far background. This might be the last image shot on film stock by Lee for Hou. Indeed, Lee might consistently shoot digitally from now on, no doubt finding properties in the digital format that can visually enhance his images. But, I for one will miss the delicate lighting of interiors in *Flowers of Shanghai* and the extreme long shot exterior landscapes (including their remote human figures) of *Puppetmaster*, that I was able to show my students on 35mm prints over ten years ago, the grainy details of which can only be fully appreciated projected on a big screen, from a film strip source.

In a long interview with Liao Ching-sung, the editor on almost every one of Hou's films, nothing was said about cutting long takes during the editing process, but Liao cites traditions of Chinese poetry and the presence of the landscape being key to making emotional transitions between shots.[64] When Lee began to work as Hou's cinematographer, the director was already recognized as being the leading figure in having developed a distinctive long take, long shot style for the Taiwanese New Wave. In interviews, both Hou and Lee refer to their belief in realism and being open to chance and therefore being prone to improvisation. As discussed here, the incidence of camera movement in Hou's films' shots suddenly shifted from an average of 27 percent in eight films (from *The Sandwich Man* in 1983 to *The Puppetmaster* in 1993) to 72 percent in *Good Men, Good Women* and then 80 percent in *Goodbye, South Goodbye*, and although Hou must have been instrumental in adopting this stylistic shift, Lee was more than just a mere collaborator.[65] One could understand this and other stylistic changes in the director's work to be going against his artistic norms, as both his practice and its critical appreciation became more international. But, conversely one could also argue that, progressively, his oeuvre was becoming stylistically more "Chinese" or "Asian."[66] In this respect, Hou's interests were drawing closer to those of Lee and editor Liao, and, certainly *The Assassin* represents the film work that most clearly reflects the move backward in time to Chinese traditions of painting, poetry and martial arts literature. As for the diversification of subject matter, both historically and geographically to contemporary Japan (*Café Lumière*) and France (*Red*

Balloon), I have attempted to argue that Hou benefitted greatly from Lee's floating flexibility in adapting camera style to content.

In a 2011 Golden Horse poll of Chinese film "experts," Hou emerged as the "greatest" Chinese-language filmmaker of all time with three films in the top-ten and another three in the top-50, but all of these films were made in the 1980s, and only one other of his films, *Flowers of Shanghai* placed in the top-100 (83rd).[67] Clearly, the Chinese film "experts" are less enamored of the "international" and much-changed Hou than American and French critics, who seem to support his work, more than ever. In a *Village Voice* poll of film critics, Hou placed #1 as the Best Director of the 1990s, while in the *Cahiers du Cinéma* poll of their critics, *Goodbye, South Goodbye* was voted the #1 film of the decade. Although Hou's films are invariably released in France, the U.S. distributors have, until recently, been reluctant to pick up his films, and in the 2001 *Voice* poll, *Millenium Mambo* was the second "Best Undistributed Film" (1996–2001), while *Cahiers* critics placed it as the third best (distributed) film of 2001. *Café Lumière*, in 2004 and *Three Times*, in 2005 proceeded to be named the "Best Undistributed" films of their years by the *Village Voice*. In 2006, the online journal, *Indiewire* took over polling North American critics from the *Voice*, and sensationally, Hou's next film, *Le voyage du ballon rouge* (Flight of the Red Balloon, 2007) won both Best Film and Best Director accolades from the 105 participating critics (43 of whom placed the film in their top tens).[68] Increasingly, Lee's work both for Hou and for other directors is getting greater recognition. In 2007, he was given the Honorary Silver Mirror award by the Films From the South Festival in Oslo, Norway, the first, (and so far, only) non-director to be so awarded. He has won the Asia Pacific Screen, Asian Film, Golden Horse and International Cinephile Society awards for Best Cinematography on *The Assassin*, while polling second in the United States, National Society of Film Critics awards and third in the *Indiewire* Critics' Poll. Most recently he won the "Outstanding Artistic Contribution" prize at the Berlin International Film Festival for the cinematography on *Chang jiang tu* (Crosscurrent, China, 2016), a film that has been championed for its relationships to Chinese landscape painting and poetry, and, as of this writing, Lee is being honored with a major retrospective at New York's MoMA.[69] May the highly creative collaboration between director and cinematographer continue well into the (digital) future.

NOTES

1. Finally, a book has been published on the collaborative nature of director/cinematographer teams: Christopher Beach, *A Hidden History of Film Style: Cinematographers, Directors, and the Collaborative Process* (Oakland: University of California Press, 2015).

2. Sven Nykvist, "The Cinematographer's View: The Director of Photography," in *Making Pictures: A Century of European Cinematography*, ed. Michael Leitch (New York: Harry N. Abrams, Inc., 2003), p. 10.

3. Claudia Monterde, *Under the Mexican Sky: Gabriel Figueroa, Art and Film* (Pasadena, California: Televisa Foundation, 2013). Other titles include, *Artes de Mexico 2: El Arte de Gabriel Figueroa*, bilingual 2nd edition (1992 [1988]); *Gabriel Figueroa: Luna Córnea*, bilingual edition (Mexico City: Editorial RM, 2009); Gabriel Figueroa, *Gabriel Figueroa y la pintura mexicana* (Mexico City: Consejo Nacional Para La Cultura y Las Artes, 1996); Elena Poniatowska, *Todo México, Tomo III: La mirada que limpia: Gabriel Figueroa* (Mexico City: Editorial Diana, 1996); Ceri Higgins, *Gabriel Figueroa: Nuevas perspectivas* (Mexico City: Consejo Nacional Para La Cultura y Las Artes, 2008); José Leyva and Mariana Riva Palacio, *El cine en México y el Mundo: Cronología de Gabriel Figueroa* (Mexico City: Televisa Foundation, 2007).

4. Peter Rist, "Visual Style in the Early Films of John Ford: From *Straight Shooting* to the end of the Fox contract in 1934" (Ph.D. diss., New York University, 1988).

5. David Bordwell, *Figures Traced in Light: On Cinematic Staging* (Berkeley: University of California Press, 2005), p 192, p. 193.

6. James Udden, *No Man an Island: The Cinema of Hou Hsiao-hsien* (Hong Kong: Hong Kong University Press, 2009), p. 46. Udden also shows that the average shot length (ASL) of Hou's first three films—11.8 seconds—is three seconds longer than the ASL of a sample of 25 Taiwanese films made between 1980 and 1982.

7. Richard I. Suchenski, "Experiments with Realism: An Interview with Chen Huai-en," in Richard I. Suchenski, ed., *Hou Hsiao-hsien* (Vienna: Austrian Film Museum, 2014), p. 201.

8. Ibid.

9. This and much of the information on Lee has been gleaned from the feature-length documentary film, *Let the Wind Carry Me: The Fleeting Moments of Mark Lee* (Taiwan, 2009), directed by Chiang Hsiu-chiung and Kwan Pung-leung.

10. Lee Daw-ming, "Wang Tung," in *Historical Dictionary of Taiwan Cinema* (Lanham, MD: The Scarecrow Press, 2013), p. 393.

11. An interview conducted by David Bordwell, Taipei Film Festival, November 1999, in Bordwell, op. cit., p. 203.

12. Udden, pp. 178–81.

13. Ibid., p. 60.

14. Emilie Yueh-yu Yeh and Darrell William Davis, "Trisecting Taiwan Cinema with Hou Hsiao-hsien," in *Taiwan Film Directors: A Treasure Island* (New York: Columbia University Press, 2005), p. 158, p. 159. Udden as well as Yeh and Davis cite David Der-wei Wang, *Fictional Realism in Twentieth Century China: Mao Dun, Lao She, Shen Congwen* (New York: Columbia University Press, 1992), and Jeff Kinkley, *The Odyssey of Shen Congwen* (Stanford: Stanford University Press, 1987), as well as Shen's autobiography and a number of interviews.

15. Yeh and Davis, p. 160.

16. Barry Salt, *Film Style and Technology: History and Analysis*, 2nd ed. (London: Starword, 1992).

17. On the homepage of the *Cinemetrics* website, there is an introduction to the "Movie Measurement and Study Tool database, www.cinemetrics.lv

18. Indeed, both James Udden and myself find statistical analysis of films to be very helpful. But, for camera movement, it could more useful to measure the actual length of time that the camera moves around its own axes (panning and tilting) and through space (tracking, dollying and craning) rather than strictly follow the Salt model in counting the number of shots in which the camera moves. But, of course, this is much more difficult to do! We also tend not to include shots that only contain very slight reframing pans or tilts.

19. Bordwell, p. 167.

20. Noël Burch, *To the Distant Observer: Form and Meaning in the Japanese Cinema* (London: Scolar Press, 1979), pp. 160–61.

21. Udden, p. 48, quoting the Chinese-language publication, Chu Tian-wen and Wu Nien-jen, *Dust in the Wind* (Taipei: Yuanliu, 1992), p. 43.

22. See, David Bordwell, "Appendix: Some Quantitative Aspects of Ozu's Films," in *Ozu and the Poetics of Cinema* (Princeton, New Jersey: Princeton University Press, 1988), p.377; K. van den Heuvel's submission to *Cinemetrics* on December 13, 2009, <http://www.cinemetrics.lv/movie.php?movie_ID=4727>

23. Richard I. Suchenski, "Exploring Visual Ideas: An Interview with Mark Lee (Ping-bin)," in Suchenski, op. cit., p.209.

24. Udden, p. 76.

25. My calculations of Udden's statistics are based on his writing that "35 [shots] have no people at all . . . ," 38 "are done in extreme long shot . . . ," and "31 other shots which are primarily long shot . . ." (Udden, p. 78). I assume he also performed his shot-by-shot breakdowns using DVDs rather than 16mm or 35mm film prints.

26. Ibid.

27. Interview with Lee, in Suchenski, op. cit., p. 209.

28. See David Bordwell, Kristin Thompson and Jeff Smith, *Film Art: An Introduction*, 11th ed. (New York: McGraw-Hill Education, 2017); for example, for "classical" Hollywood films: "Typically the plot focuses on one or two central characters who want something. Characters' desires set up a *goal*, and the course of the narrative's development will most likely involve the process of achieving that goal . . . Classical Hollywood filmmakers tend to let psychological causes motivate most events. Throughout, motivation in the classical narrative film strives to be as clear and complete as possible . . ." p. 98; and for Italian neorealism, "reacting against the intricate plots found in light popular cinema, the Neorealists tended to loosen up narrative relations," where "the most formally innovative Neorealist films allow the intrusion of scenes that aren't motivated causally—that seem, in fact, to be accidents." p. 478, and, [Roberto] Rossellini's *Paisan* [1946] is episodic, presenting six anecdotes of life in Italy during the Allied invasion. Often, we are not told the outcome of an event, the consequence of a cause." p. 479.

29. In a more recent analysis where he claims that *Dust in the Wind* is both Hou's "definitive" film, and "of the New Cinema as a whole," James Udden writes that "In a sense, what Hou offers the viewer is not cause and effect, but rather effects almost drowned out by the endless details of every day life. As a result, the causes are left for us to surmise. His extremely elliptical and oblique narration is risky to say the least,

and yet what holds it all together is a complex weave of motifs, all of resonate with each other." "*Dust in the Wind*: A Definitive Hou/New Cinema Work," in Suchenski, op. cit., p. 132, p. 139.

30. Udden, (2009), p. 82.

31. Haden Guest, "Reflections on the Screen: Hou Hsiao Hsien's *Dust in the Wind* and the Rhythms of the Taiwan New Cinema," in Chris Berry and Feii Lu, ed., *Island on the Edge: Taiwan New Cinema and After* (Hong Kong: Hong Kong University Press, 2005), p. 28, p. 30.

32. Interview with Lee, in Suchenski, op. cit., p. 209.

33. Ibid.

34. In writing about Hou for a New York retrospective in 2014, J. Hoberman says he "first became convinced of his genius when [he] saw *The Puppetmaster* (1993) some twenty years ago." He concludes his article by returning to this film as the work for which "Hou first developed a startlingly advanced form of montage that has been compared to the movement of clouds drifting across the sky" and by asking the question, "Is there another filmmaker who can so fluidly celebrate the moment as well as the epoch, and do so in the same shot?" In *The New York Review of Books*, daily (September 19, 2014, 4:45 pm), online: http://www.nybooks.com/daily/2014/09/19/taiwan-master-timekeeper-hou-hsiao-hsien/?insrc=hpbl

James Udden (2009) concludes his chapter on the two "history" films by saying: "Without dint of logic or argumentation, without the usual trappings of intellectual fervor or sophisticated nattering, *The Puppetmaster*, by the very fabric of its cinematic being, is one of the most profound philosophical statements to ever grace the screen." p. 130.

35. Udden (2009), p. 120, p. 180.

36. Interview with Lee, in Suchenski, op. cit., p. 210.

37. Udden (2009), p. 122

38. See Sylvia Li-Chun Lin, *Representing Atrocity in Taiwan: The 2/28 Incident and White Terror in Fiction and Film* (New York: Columbia University Press, 2007), p. 54. The camera operator for *Good Men, Good Women* is credited as Han Yong-chon. In an interview with Udden (May 1, 2001), Chen suggested that the inclusion of camera movements in *Good Men* could be attributed to the uncertain work of a new operator (2009, p. 138).

39. Interview with Lee, in Suchenski, op. cit., pp. 211–12.

40. Hasumi Shigehiko, "Who Can Put Out the Flame?": On Hou Hsiao-hsien's *Flowers of Shanghai*, in Suchenski, op. cit., p. 111.

41. Lee responding to my questions sent to him via Chang Chuti, an assistant to Hou and Lee, on May 10, 2016. His replies were sent by Ms, Chang on May 31, and translated by (Hannah) Pan Weixian. In his reply to the *Flowers of Shanghai* question he noted that some of the sets were built on location rather than in a studio.

42. Udden (2009), p. 149.

43. Hou, Interview with Li Dayi, "Hou Hsiao-hsien's Cinematic Life," *Film Appreciation*, 17, no. 3 (May/June, 199):82, in Udden (2009), p. 148.

44. Interview with Lee, in Suchenski, op. cit., pp. 212–13.

45. Udden (2009), pp. 150–52.

46. See Steven Teo, *Wong Kar-Wai* (London: British Film Institute, 2005), p. 128, where, in addition to saying that Lee "took over after Chris Doyle shot about a third

of the film," he stated that "the colors . . . are much more refined and subdued, unlike the hyped colours of *Happy Together*."

47. In a medium-length documentary made by Wong, *@In the Mood for Love* included on Disc 2 of the Criterion Collection DVD set in 2002, one can glimpse Lee on a camera dolly running along tracks to follow Leung walking along the sidewalk.

48. Statistics are taken from Udden (2009), p. 181.

49. Bordwell (2005), p. 186.

50. Ibid., 235–36.

51. Chang interview with Lee, op. cit. On lenses, Lee said, "I prefer lenses with wider apertures because I like genuine, realistic light. I want to capture the delicate available light and color from a scene and preserve them as much as possible. A wider-aperture lens is easier to master too. It produces a more shallow depth of field and keeps the theme of a scene in focus."

52. Abel Fan counted 95 shots, with an ASL of 68.6 secs; I think he is wrong in recognizing that only 70 of them contain camera movement; for *Cinemetrics* http://www.cinemetrics.lv/movie.php?movie_ID=1973

53. My analysis here is derived in part from the "making of" documentary included as a special feature on the Palm Pictures DVD, MVD 1377 (2002).

54. I realized how long the film crew of *Café Lumière* would have to wait for such an unlikely occurrence when my friend Randolph Jordan and I, after finding the exact spot, spent hours on two separate days unsuccessfully trying to capture the coincidence of three trains crossing at exactly the same time on camera.

55. Udden (2009), p. 180.

56. See Udden, in Davis and Chen, op. cit., p. 180. See also, Song Wee Lim, *Tsai Ming-Liang and a Cinema of Slowness* (Honolulu: University of Hawai'i Press, 2014), p. 88 for a comparison of Hou and Tsai's ASL and camera movement statistics.

57. David Bordwell and Kristin Thompson (2017), pp. 466–67. For definitions of Impressionist paintings, see John I. Clancy, ed., *Impressionism: Historical Overview and Bibliography* (New York: Nova Science Publishers, Inc., 2003), p.1. "The primary concern of the Impressionists was the effect of light on an object rather than the exact presentation of the form. . . . Part of the wide appeal of Impressionist paintings may be the fact that the human brain tends to generally perceive objects using what might be called an impressionist filter. . . . It was the novel use of color which influenced many artists of later generations and movements and which established Impressionism as a seminal movement in the history of art." See also, Rachel Barnes, ed., *Renoir by Renoir: Artists by Themselves* (New York: Alfred A. Knopf, 1990), p. 8. "In the summer of 1869 the two painters [Auguste Renoir and Claude Monet] worked together at Grenoillère, on the River Seine. Their shared fascination with working in the open air and studying the transient effects of light, atmosphere and weather conditions on the surface of the water produced the first revolutionary Impressionist works." For Lee's cinematography, "glass" could replace "water."

58. There are three nominations for the ASC Spotlight award, which was inaugurated in 2014 to "recognize outstanding cinematography in features and documentaries that are typically screened at film festivals, in limited theatrical release, or outside the United States. ASC members submit entries for

consideration, and a blue-ribbon panel chooses the nominees. All active ASC members may vote to select the winner." http://www.theasc.com/site/news/asc-nominates-flinckenberg-khondji-landin-for-spotlight-award/

59. See interview with Aliza Ma, "Killer Technique," in *Film Comment*, 51, no. 5 (September/October 2015): 29. "I specifically wanted to use a wind-up. You wind the camera up, and shoot for less than one minute, and you have to wind it up again. I like the inscribed structure of this filming process and the inherent limitation it posed. I always want to set some kind of external limit for myself before working. . . . I believe that imposing such limits forces an artist to be more creative."

60. Much of my analysis here is based on conversations I had with Toronto's Chinese film expert, Shelly Kraicer and Li Cheuk-to, the Artistic Director of the Hong Kong International Film Festival Society at the Toronto International Film Festival on September 10, 2015.

61. Interview with Ma, p. 33.

62. From an e-mail exchange, translated by La Frances Hui, in Daniel Eagan, "Master of Light: Museum of Modern Art celebrates DP mark Lee Ping-Bing," *Film Journal International*, June 14, 2016: http://www.filmjournal.com/features/master-light-museum-modern-art-celebrates-dp-mark-lee-ping-bing

63. Richard I. Suchenski, "Working within Limits: A Conversation with Hou Hsiao-hsien and Chu Tien-wen," in Suchenski, ed., op cit., pp. 193–94.

64. Richard I. Suchenski, "Finding the Right balance: an Interview with Liao Ching-sung," with assistance from Chang Chuti, trans. Dennis Li, in Suchenski, ed., op. cit., pp. 215–21. Liao said that poets of the Tang Song and Yuan dynasties "would project those emotions onto objects and the landscape. You can recognize these traits in the very poetic films of Hou. Since he tends to use amateur actors, he uses long shots to keep the camera away from them. One effect of that is that the focus is not on the expressions of the performers, but on the atmosphere created by objects and the landscape. There is an ambience to these that matches up with the tradition of Chinese poetry, particularly in the shifts from long shots to closer shots." p. 217.

65. See James Udden, "This Time He Moves!: the deeper significance of Hou Hsiao-hsien's radical break in *Good Men, Good Women*," in Darrell William Davis and Ru-shou Robert Chen, ed., *Cinema Taiwan: Politics, Popularity and State of the Arts* (London: Routledge, 2007), pp. 200–201.

66. See, for example, Richard I. Suchenski, "Also Like Life: The Films of Hou Hsiao-hsien," the introduction to his edited book, especially, where he writes; "In different, but complementary ways, the essays by [Peggy] Chiao, [Jean-Michel] Frodon, and Ni Zhen consider Hou's evolving interest in Chinese aesthetics. Chiao and Ni help to explicate Hou's debt to Chinese classics he studied in his youth, while Frodon's arguments—explicitly influenced by the work of French Sinologist François Julien—relate his work to developments in postwar international cinema as well as premodern Chinese culture." Suchenski, op. cit., p. 22.

67. See, Vey Lo, Vicky Lee, Wei I-Ying, and Chiamin Liu, eds., *The River of Shadow and Light: Greatest Chinese-Language Films* (Taipei, Taiwan: Golden City Publishers, 2011). The eventual top-100 list does seem somewhat biased toward Taiwanese films, with seven features appearing in the top ten, pp. 420–23.

68. The *Village Voice* continued their poll, but in firing Dennis Lim, who started the poll and then other critics including the irreplaceable J. Hoberman in 2012, the famous New York weekly newspaper has gradually lost its credibility. Most of the poll results can no longer be accessed, but for the "iW Critics Poll '08" see http://www.indiewire.com/article/complete_results

69. The series opened with *Crosscurrent* on June 16, 2016, and continued with four films directed by Hou, Tian's *Springtime in a Small Town*, the documentary, *Let The Wind Carry Me*, and Wong Kar-wai's *In the Mood for Love* that are discussed here, as well as a film by Ann Hui, Bourdos' *Renoir* and Tran's *The Vertical Ray of the Sun* and *Norwegian Wood* all of which are mentioned here.

BIBLIOGRAPHY

American Society of Cinematographers (ASC), online. <http://www.theasc.com/site/news/asc-nominates-flinckenberg-khondji-landin-for-spotlight-award/>

Barnes, Rachel, ed. *Renoir by Renoir: Artists by Themselves*. New York: Alfred A. Knopf, 1990.

Beach, Christopher. *A Hidden History of Film Style: Cinematographers, Directors, and the Collaborative Process*. Oakland: University of California Press, 2015.

Bordwell, David. *Figures Traced in Light: On Cinematic Staging*. Berkeley: University of California Press, 2005.

———. *Ozu and the Poetics of Cinema*. Princeton, New Jersey: Princeton University Press, 1988.

Bordwell, David, Kristin Thompson and Jeff Smith. *Film Art: An Introduction*, 11th ed. New York: McGraw-Hill Education, 2017.

Burch, Noël. *To the Distant Observer: Form and Meaning in the Japanese Cinema*. London: Scolar Press, 1979.

Chen Huai-en. "Experiments with Realism: An Interview with Chen Huai-en." Interview by Richard I. Suchenski. In *Hou Hsiao-hsien*, ed. Richard I. Suchenski. 199–208. Vienna: Austrian Film Museum, 2014.

Cinemetrics. "Movie Measurement and Study Tool," database, online. <www.cinemetrics.lv>

Clancy, John I., ed. *Impressionism: Historical Overview and Bibliography*. New York: Nova Science Publishers, Inc., 2003.

Eagan, Daniel. "Master of Light: Museum of Modern Art celebrates DP mark Lee Ping-Bing." *Film Journal International*, June 14, 2016, online: <http://www.filmjournal.com/features/master-light-museum-modern-art-celebrates-dp-mark-lee-ping-bing>

Guest, Haden. "Reflections on the Screen: Hou Hsiao Hsien's *Dust in the Wind* and the Rhythms of the Taiwan New Cinema." In *Island on the Edge: Taiwan New Cinema and After*, ed. Chris Berry and Feii Lu. 27–37. Hong Kong: Hong Kong University Press, 2005.

Hoberman, J. "Taiwan Master Timekeeper Hou Hsiao-hsien." *The New York Review of Books*, daily (September 19, 2014, 4:45 pm), online: <http://www.nybooks.com/daily/2014/09/19/taiwan-master-timekeeper-hou-hsiao-hsien/?insrc=hpbl>

Hou Hsiao-hsien and Chu Tien-wen. "Working within Limits: A Conversation with Hou Hsiao-hsien and Chu Tien-wen" Interview with Richard I. Suchenski. In *Hou Hsiao-hsien*, ed. Richard I. Suchenski. 186–98. Vienna: Austrian Film Museum, 2014.

"iW Critics Poll '08." *Indiewire*, online: <http://www.indiewire.com/article/complete_results>

Lee Daw-ming. "Wang Tung," In *Historical Dictionary of Taiwan Cinema*. Lanham, MD: The Scarecrow Press, 2013.

Lee Ping-bin, Mark. "Exploring Visual Ideas: An Interview with Mark Lee (Ping-bin)." Interview with Richard I. Suchenski. In *Hou Hsiao-hsien*, ed. Richard I. Suchenski. 209–14. Vienna: Austrian Film Museum, 2014.

Let the Wind Carry Me: The Fleeting Moments of Mark Lee. Directed by Chiang Hsiu-chiung and Kwan Pung-leung. 86 min. Taiwan, 2009. Film, viewed online, vimeo.com.

Liao Ching-sung. "Finding the Right balance: an Interview with Liao Ching-sung." Interview with Richard I. Suchenski, with assistance from Chang Chuti, trans. Dennis Li. In *Hou Hsiao-hsien*, ed. Richard I. Suchenski. 215–21. Vienna: Austrian Film Museum, 2014.

Lim, Song Wee. *Tsai Ming-Liang and a Cinema of Slowness*. Honolulu: University of Hawai'i Press, 2014.

Lo, Vey, Vicky Lee, Wei I-Ying, and Chiamin Liu, ed. *The River of Shadow and Light: Greatest Chinese-Language Films*. Taipei, Taiwan: Golden City Publishers, 2011.

Ma, Aliza. "Killer Technique." *Film Comment* 51, no. 5 (September/October 2015): 29–33.

Monterde, Claudia. *Under the Mexican Sky: Gabriel Figueroa, Art and Film*. Pasadena, California: Televisa Foundation, 2013.

Nykvist, Sven. "The Cinematographer's View: The Director of Photography." In *Making Pictures: A Century of European Cinematography*, ed. Michael Leitch, 10. New York: Harry N. Abrams, Inc., 2003.

Rist, Peter. "Visual Style in the Early Films of John Ford: From *Straight Shooting* to the end of the Fox contract in 1934." Ph.D. diss., New York University, 1988.

Salt. Barry. *Film Style and Technology: History and Analysis*, 2nd ed. London: Starword, 1992.

Shigehiko, Hasumi. "Who Can Put Out the Flame?: On Hou Hsiao-hsien's *Flowers of Shanghai*." In *Hou Hsiao-hsien*, ed. Richard I. Suchenski. 106–17. Vienna: Austrian Film Museum, 2014.

Suchenski, Richard I. "Also Like Life: The Films of Hou Hsiao-hsien." In *Hou Hsiao-hsien*, ed. Suchenski. Vienna: Austrian Film Museum, 2014.

Teo, Steven. *Wong Kar-Wai*. London: British Film Institute, 2005.

Udden, James. "*Dust in the Wind*: A Definitive Hou/New Cinema Work," In *Hou Hsiao-hsien*, ed. Richard I. Suchenski. 130–43. Vienna: Austrian Film Museum, 2014.

———. *No Man an Island: The Cinema of Hou Hsiao-hsien*. Hong Kong: Hong Kong University Press, 2009.

———. "This Time He Moves!: the deeper significance of Hou Hsiao-hsien's radical break in *Good Men, Good Women*." In *Cinema Taiwan: Politics, Popularity and*

State of the Arts, ed. Darrell William Davis and Ru-shou Robert Chen. 183–202. London: Routledge, 2007.

Yeh, Emilie Yueh-yu and Darrell William Davis. "Trisecting Taiwan Cinema with Hou Hsiao-hsien." In *Taiwan Film Directors: A Treasure Island.* New York: Columbia University Press, 2005.

Chapter 7

The Making of Hong Kong

The Texture of Cinema

Carlos Rojas

The Hong Kong director Fruit Chan 陳果 once remarked, "I don't like DV [digital video], because I grew up in the age of film. Nevertheless, I cannot deny technology's advance. Given this rapid and extraordinary technological advance, you have no choice but to use DV." At the time he said this, in early 2002, Chan was finishing his first digital film, *Public Toilet* (人民公廁), and he recalled how

> at first, I found it hard to get used to DV. I felt it was much like television, and found that it had a very different texture from film. It didn't feel very comfortable or natural, and consequently I even stopped filming *Public Toilet* altogether for half a year. However, I gradually discovered that DV was actually quite convenient. We could shoot on US streets as though we were making a home documentary. This kind of freedom was really quite nice."[1]

Chan here not only emphasizes film's unique "texture," he also notes that DV allows him to film half a world away as though he were in his own home. For Chan, accordingly, film connotes a sort of material specificity wherein the conditions of production remain tangible within the finished product, while DV is associated with a process of abstraction that permits the user to bridge vast geographical distances. Whereas the indexical quality of film offers a direct link between the cinematic representations and the original visual scenes that produced it, the link between reality and representation in digital media is much more mediated, as visual stimuli are first translated into data before being translated back into visual images.

In the following discussion, I will examine a pair of films from two different pivotal moments in Chan's career—one being his first independent film, *Made in Hong Kong* (香港製造), from 1997, and the second being his first

DV film, *Public Toilet*, from five years later—and will consider how each work deploys some of the distinctive qualities of its own cinematic medium while at the same time offering a commentary on the national and transnational position of Hong Kong itself.

MADE IN HONG KONG

Fruit Chan's emphasis on film's distinctive texture in these 2002 remarks is appropriate, given that his first break-out work five years earlier used not only film but more specifically old film stock that exaggerated the medium's material properties. After Chan's directorial debut, *Finale in Blood* (大鬧廣昌隆)—which was completed in 1991, but was not released until 1993—performed poorly at the box office, Chan had to wait several years before having another opportunity to direct a film of his own. He ultimately decided to produce an ultra low-budget production, for which he wrote the script, saved up about 500,000 HK dollars (approximately US$60,000), acquired over 80,000 feet of leftover film stock, and hired several non-professional actors. The result was *Made in Hong Kong*, which opened in 1997 and went on to win a number of major awards, including Best Picture at the 1998 Hong Kong Film Awards. The film also occupies an important position within Chan's cinematic career, in that it not only became the first installment in the director's critically acclaimed Handover trilogy, in which Chan explores the political and sociocultural implications of Hong Kong's 1997 return to Chinese control, but furthermore the budgetary restrictions under which the film was produced helped generate a distinctive visual aesthetic—and specifically an elegantly "unpolished feel"—that Chan would retain in many of his subsequent works, even after they were not subject to the same logistical constraints.

After Chan finished filming *Finale in Blood* in 1991, he spent several years assisting on other films, including David Li's 1994 Hong Kong mafia film *Heaven and Earth* (天與地). After *Heaven and Earth* was completed, Chan noticed that the studio still had about 40,000 feet of unused film stock and short ends (the remainders of partially used reels), and the film's producer, Andy Lau, allowed Chan to keep this leftover film stock while also arranging for his production company, Team Work Production House, to supply Chan with an additional forty thousand feet. The film stock in question was old—Chan observes at one point that the *Heaven and Earth* remainders were "almost expired" by the time he started using them,[2] while his cinematographer O Sing-pui recalls that much of the stock was already more than seven years past its expiration date.[3] Footage shot on aged film stock typically has a faded, grainy appearance, and distorted coloration, and *Made in*

Hong Kong features many of these same visual qualities—be they a result of the film stock that was being used and/or as a deliberate technique employed by the director and his cinematographers (O Sing-pui and Lam Wah-Chuen).

Although at one level the film's resulting visual aesthetic is a practical consequence of the budgetary constraints under which Chan was working, at another level this aesthetic nicely complements the content of the work itself. For instance, this aesthetic lends the film a distinct independent documentary feel, which is appropriate given the work's focus on the private lives of youth from Hong Kong's underclass—youth who would be nearly invisible within conventional representations of the city. Independent documentaries are typically produced on modest budgets, and therefore frequently make many of the same sorts of compromises that Chan made for *Made in Hong Kong*. In addition, independent documentary filmmakers also often chose to exaggerate the rough and unpolished appearance of their final product, so as to make it appear less crafted and consequently more "real." The irony, of course, is that many of these same techniques not only reinforce the presumed reality of the subject matter being depicted, they also underscore the constructed nature of the documentary itself.

The stark budgetary constraints under which Chan's film was produced also resonate with the work's subject matter, as *Made in Hong Kong* focuses on three teenagers who all live in poor neighborhoods of Hong Kong. The film's protagonist is a rail-thin high school dropout by the name of Autumn Moon (played by Sam Lee), who collects debts from other low-income residents on behalf of a loan shark affiliated with the triads (the Hong Kong mafia). Moon is also the friend and protector of Sylvester (Wenbers Li Tung-cheun), a mentally handicapped young man constantly being bullied by local youth, who pursue him in packs and beat him savagely for no apparent reason. One day when Moon and Sylvester are attempting to collect a debt from a woman living in a low-income housing unit, they happen to catch a glimpse of the woman's sixteen-year-old daughter—an attractive short-haired girl by the name of Ping (Neiky Yim Hui-Chi)—and are both immediately stunned by her beauty.

The three youngsters become close friends, forming a virtual family that comes to stand in for their respective fractured families, given that each of the three youngsters has been abandoned by either one or both of their parents. For instance, Moon and his mother have been abandoned by his father (who has taken a mistress, with whom he has had a young child), and by the end of the film Moon's mother has left him as well. Meanwhile, Sylvester has been separated from his entire family, and now lives under Moon's informal supervision. Finally, Ping, lives alone with her mother, who is constantly hounded by debt collectors attempting to collect money owed by her estranged husband. In fact, Moon first meets Ping while attempting to collect on her father's debts, but after Moon gets to know her he begins working on her behalf—attempting to give Ping and her mother money while defending them from other debt collectors. Each of these three youngsters, accordingly, are presented as fractured elements forgotten by society and abandoned by their own families, but which are woven together into new social units—much the same way that Fruit Chan took film stock left over from earlier productions and spliced it together to yield *Made in Hong Kong* itself.

In the film, Moon's work collecting small debts from other low-income residents on behalf of local triads stands in stark contrast not only to the glitzy world of finance capital for which Hong Kong is best known in real life, but also to the glamorous representations of triad life for which Hong Kong is perhaps best known within popular culture. The film's recycling and transformation of gangster imagery can been observed at a number of different levels. First, *Heaven and Earth*, the 1994 triad film from which Chan obtained much of the film stock for *Made in Hong Kong*, was itself a mafia film, and was specifically a remake of the Brian De Palma's iconic 1994 gangster film *The Untouchables*. Fruit Chan's own *Made in Hong Kong*, meanwhile, not only tropes off of *Heaven and Earth* but furthermore may be viewed as a second-order adaptation of De Palma's earlier film in which it is based. Moreover, Andy Lau, the star and producer of *Heaven and Earth*, also starred in and produced the popular 1990s Hong Kong gangster series *The Young and Dangerous*, which is also an important source of inspiration for *Made in Hong Kong*.[4] Within Fruit Chan's film itself, meanwhile, Moon's bedroom is decorated with movie posters for films such as Oliver Stone's *Natural Born Killers* (1994) and Luc Besson's *Leon the Professional* (1994), both of which offer a glamorized depiction the sort of gangster-style violence on which Chan's film is modeled, even as it simultaneously offers an ironic commentary on the genre itself.

The influence, in *Made in Hong* Kong, of these sorts of gangster and mafia films can be seen particularly clearly in a sequence that begins when Moon's triad boss gives him a fake handgun, and Moon proceeds to jump around the office while mimicking iconic poses from mainstream gangster films. Later, after Moon is given a real gun and instructed to shoot two businessmen on behalf of his triad boss, we see Moon dancing around his bedroom while holding his gun. With loud techno music playing in the background, we see shots of Moon dancing are interspersed with silhouette shots of a handgun against a white backdrop (as in the credit sequence to a James Bond film), and shots of the *Natural Born Killers* poster on his bedroom wall.

When it comes time to carry out the actual assassination, however, Moon discovers that he is unable to follow through with it, and the result is an ironic parody of the gangster films on which he is implicitly modeling himself. In fact, we see short fragments of the assassination scene several times in succession. First, in the initial sequence we see what appears to be Moon's own fantasy of how the event will unfold (the first sequence is shot entirely in slow motion, as we see Moon run up to two businessmen and point his gun at them, then there is the sound of two gunshots just as the camera cuts away, after which we see Moon sprinting down a hallway and deftly tossing the gun into a trash can). In subsequent iterations, meanwhile, we see a much different vision of the encounter, which presumably accords more closely with what actually occurred (in particular, we see Moon run up to the businessmen and raise his gun, but then quickly lower it again and run away without firing a shot, after which we see him rushing frantically down the hallway and tripping over the trash can instead of deftly tossing his gun into it). In this way, Chan's film tropes on some of the cinematic conventions of Hong Kong gangster films (including the use of slow-motion, freeze frames, and a montage of very short takes from different perspectives), while at the same time tacitly drawing attention to the film's own cinematic medium (by deliberately foregrounding a set of easily recognizable cinematic editing techniques).

It turns out that Moon's decision to accept the assignment to kill the businessmen was driven not by a desire to take life, but rather to save it. In particular, the assassination attempt is preceded by a sequence in which we see Moon rolling himself a marijuana joint and languidly smoking it—as his voice-over explains that he was actually extremely nervous about the assignment, but had accepted it only to help Ping. Shortly after initially meeting Ping, Moon had learned that the teenager was dying of acute renal failure, whereupon he became increasingly desperate to find a way to try to help her. At one point he even tried to give her one of his own kidneys, but ran into bureaucratic difficulties. His narrative voice-over preceding the assassination sequence, accordingly, emphasizes the fact that he accepted the assignment specifically in order to earn money for Ping's medical expenses. Even incentivized by this noble motivation, he is still unable to carry through with the plan and take the lives of the two strangers. Here, Moon's inability to be the sort of cold-blooded killer found in gangster films comes full circle with his determination to find a way to save Ping's life.

Made in Hong Kong is haunted not only by the anticipation of Ping's death, but also by the death of the film's fourth main character: a schoolgirl by the name of Susan Hui (played by Amy Tam Ka-cheun). We first see Susan near the beginning of the film—in a scene shot in an eerie blue light—sitting on the roof of a building, but a few seconds later she gets up, walks to the end of the building, and jumps off. Her body happens to land right in front of Sylvester, who retrieves two blood-soaked letters she was holding when she jumped. He then immediately runs off, pursued by the teenage bullies who terrorize him throughout the film, and when the bullies catch him they beat him so badly that he ends up in the hospital. In the hospital, a physician retrieves the two letters from Sylvester's pocket and hands them over to Moon, who has come to fetch his friend. It turns out that the letters are a pair of suicide notes, one of which is addressed to Susan's parents and the other is addressed to the high school teacher with whom Susan had had an affair, and on account of whom she had decided to kill herself.

The moment Moon receives the letters in the hospital, the film immediately cuts to another shot of Susan jumping to her death—with this scene appearing in the same eerie blue light as the first, but seen from a slightly different angle. It is unclear whether these sequences present the suicide as it actually happened or as Moon comes to imagine it, though visual elements from these first two depictions of Susan's death recur repeatedly in a series of erotically-charged dreams that Moon has throughout the film. Moon has the first of these dreams shortly after he returns home from the hospital. In the dream sequence, which is also shot in a blue light, we see Susan fall and hit the pavement, whereupon a white liquid begins gushing out from beneath her body. A few seconds later, we see the same scene again, but this time with dark red blood where the white liquid had been. Eventually, Moon wakes up covered in sweat, and immediately reaches discretely for his crotch.

At a thematic level, the white liquid gushing out from beneath the dead girl's body might be viewed as a symbol of innocence, purity, or even death itself (in Chinese culture, the color white connotes death). Alternatively, it may also been seen as a symbol of the semen that Moon—in a ritual that is repeated multiple times throughout the film—immediately proceeds to rinse from his underwear after he wakes up from his dream. At a strictly visual level, meanwhile, the juxtaposition of the white and dark red liquid in these two sets of shots from Moon's dream makes the white liquid appear to be negative exposures of the red one—suggesting that Moon's erotic fantasies are fundamentally filmic in nature, with desire functioning as a negative image of the specter of death. In this way, the imbrication of the Susan's suicide and Moon's subsequent erotic fantasies mirrors the corresponding imbricated relationship between Fruit Chan's fictional film and the social realities by which it is inspired.

In making *Made in Hong Kong,* Fruit Chan saved money not only by using leftover film stock from other productions but also by using first-time, nonprofessional actors. For instance, Chan recruited Sam Lee (who plays Moon) after he found him skate-boarding in the street, and he assigned the film's own script supervisor, Li Tung-cheun, to play the part of Sylvester.[5] Not only does this reliance on nonprofessional actors reinforce the film's documentary feel, it also had practical implications for the film's structure and content. In particular, Fruit Chan explains that the original script did not include any voice-over narration, but rather the voice-overs were added after the fact "to the scenes where the acting was not competent enough."[6] Despite its origins as a practical response to the contingencies of the production process, however, this voice-over narration went on to become one of the defining elements of the film as a whole.[7]

Made in Hong Kong is framed by a pair of extradiegetic voice-overs. The work opens with a scene of an urban playground seen through a chain link fence, and after a few seconds we hear Moon introducing himself:

> My name is Autumn Moon. I quit school after junior high. I was no good in my studies. But the education system was no better. It not only excludes me from further study, it also produces juveniles like me. I've got nothing to do but hang around public playgrounds, playing basketball, taking part in brawls, and being recruited to join the triads or recruiting others to become triad members.[8]

As this initial voice-over proceeds, we see Moon playing basketball with his friends. The precise relationship here between the image on-screen and the extradiegetic voice-over is not specified, though viewers may initially infer that the voice-over articulates the inner thoughts of the protagonist, Moon.

This assumption that the voice-over corresponds directly to character we see on scene, however, is challenged at the end of the film, in a sequence in a cemetery when we hear Moon's voice-over saying,

> The living often like to pass judgment on the dead. They despise the latter for their cowardice. I wonder how many people would understand how Ping, Sylvester, Susan, and I actually think? . . . No one would. But there's one thing I'm certain of, which is that we are all very happy now. We aren't afraid even if we face an uncertain world. Because we've been immune.

This latter remarks parallel the work's opening voice-over, in that both emphasize Moon's feeling of being overlooked by mainstream society, but whereas in the first instance Moon was speaking of himself in the first-person singular ("I"), in this final instance he is explicitly including his friends Sylvester and Ping in this self-description ("we"), and even includes Susan, despite the fact that he has never met her in person. Over the course of the film, accordingly, Moon's voice-overs transition from articulations of his own perspective broader statements on behalf of the intimate social group to which he belongs.

Moreover, by the time these final words are heard, Moon, Sylvester, and Ping are all dead. Moon had been attacked by a rival gangster and ended up in the hospital in a coma, and after he eventually recovered and was released, he learned that while he was unconscious Sylvester had been killed by gangsters and Ping had also died embracing him on his hospital bed. Moon therefore resolves to avenge Sylvester and Ping by assassinating the two triad leaders who were responsible for Sylvester's death and for harassing Ping's mother for debt payment. After carrying out the assassinations, Moon proceeds to the cemetery where Ping is buried, sits down next to her tombstone, then shoots himself in the head. Moon's body sits unnoticed at Ping's grave for a couple of days, until it is finally discovered by some children who have come to the cemetery to fly a kite. It is at this point that we hear Moon's final voice-over, speaking to us from beyond the grave—which in turn raises the possibility that *all* of Moon's voice-overs in the film may originate from a postmortem realm.

In addition to Moon's monologues throughout the film, another set of voice-overs involve the two letters that Sylvester finds on Susan's dead body. Relatively early in the film, Moon, Sylvester, and Ping return the first letter to Susan's former teacher, but he rips it up (unopened) as soon as they are out of sight. Just before Moon commits suicide, meanwhile, he drops off the second letter at the home of Susan's parents, who immediately open the envelope and read the letter carefully. The camera shows a close-up of the page, revealing that Susan's note is immediately followed by a second note written by Ping and addressed to Susan, which in turn is followed by a third note written by Moon after Ping's death and addressed to both Susan and Ping. While Susan's note attempts to explain why she has decided to take her own life, Ping's note instead criticizes Susan for being insensitive and selfish, and notes that Ping herself is not afraid to die since she still has Moon. Moon's note at the bottom of the page, meanwhile, observes that he has become obsessed by both girls, and concludes by noting that at least the girls had their families, while he himself perhaps is better off dead. As the camera pans back and forth across the page, we hear each of the three youngsters reading their respective portions of the letter in their own voice—though by the time the Susan's parents receive the actual letter Susan and Ping are already dead, and Moon soon will be as well. In this way, their voices come to have a spectral quality, bridging the world of the living and that of the dead.

All three notes are written on the same blood-soaked sheet of stationary, at the bottom of which is printed the phrase, "I was awakened by a dazzling light in the morning." This phrase, which appears in English in florid red letters, is directly overlaid by the final handwritten line of Moon's own note, in which he cites Ping's mother's remark that "If we die young, we'll remain forever young." The result is a palimpsest of multiple texts and voices reflecting on youth, death, and the possibility of the afterlife. Although nearly illegible when the film is viewed in real time, the inscription at the bottom of the stationary both complements the sentiments expressed in the handwritten notes on the same page, and at the same time offers a timely reminder that the medium on which those notes are written has been borrowed from a different source—the same way that the film stock Chan used for his film was itself borrowed from other sources.

These twin themes of youth and morning light raised at the end of the suicide note are brought together in the film's final voice-over a few minutes later. Immediately following Moon's "we've been immune" remark—which accompanies the shot of his dead corpse leaning against Ping's tombstone—we hear a woman's voice saying in Cantonese, "This is your world, and it is also ours. But at the end of the end, it is yours. You youth . . . are like the morning sun, and we have placed our hopes in you." There is a brief pause and then the woman continues, "You're listening to the People's Radio of Hong Kong. What we have quoted was a speech given by Chairman Mao to the leaders of the youth. Let's repeat and study the message in *Putonghua* [Mandarin]." Just before the announcer begins re-reading the same passage in Mandarin, however, the film cuts to the closing credits.

This final voice-over, with its quotation of Mao and its proposal to transition from Cantonese to Mandarin, contains one of the film's most explicit references to Hong Kong's Handover to Mainland China. In particular, even though the Joint Agreement specified that Hong Kong would retain its current sociopolitical and economic system for at least fifty years following the Handover, there was nevertheless considerable concern in Hong Kong with how China's socialist system would meld with Hong Kong's existing capitalist system—and this fragment of a radio broadcast hints at the conflicted attitude with which many Hong Kongers viewed the impending Handover.

Resonating suggestively with the "dazzling light in the morning" phrase printed on the bottom of the stationary on which Susan writes her suicide note, moreover, the "morning sun" speech being quoted here was one that Mao gave to some Chinese foreign students studying in Moscow on November 17, 1957, precisely forty years before the 1997 setting of Chan's film. The sunny optimism of Mao's insistence here that the future belongs to the youth is contrasted not only with the actual conclusion of Chan's film (in which all four of the youthful protagonists end up dead), but also with another speech that Mao gave in Moscow the very next day. In his "American Imperialism is a Paper Tiger" speech he delivered at the Moscow Meeting of Representatives of the Communist and Workers' Parties on November 18, 1947, Mao declared that

> American imperialism is a paper tiger, and nuclear war is not frightening. There are 2.7 billion people in the world, and it is of little consequence if some of them are killed. There are 600 million people in China, and even if half of them are killed, there will still be 300 million people left. Whom should I be afraid of?[9]

Although accounts of these remarks had circulated informally for some time, their authenticity was dramatically reconfirmed in early 2013 when China's documentary channel, CCTV-9, publically broadcast a series of the recently declassified China-Russia foreign affairs files, which included Mao's nuclear war speech.[10] Just as Mao's 1957 Moscow speech offers a vision of a future that is predicated on the necessary possibility of death and destruction, it also happened to be offered on the eve of China's Great Leap Forward (1958–1961)—which sought to catapult China past Great Britain as one of the world's largest economies, but which ultimately yielded not increased prosperity but rather a nationwide famine in which as many as 30 million Chinese would lose their lives.

Just as the historical juxtaposition of Mao's famous "morning sun" remarks and his subsequent "nuclear war" discussion illustrates a crucial tension within Maoism itself, it also points to a defining thematic element in Chan's film. As an ideology, Maoism was very future-oriented, embracing the possibility of revolutionary transformation, though this quest for progress carried the necessary possibility of death and destruction. In Chan's film, meanwhile, the emphasis on youth is paired with a parallel attention to death, such that the futurity embodied by the youth is perpetually inflected with the specter of loss, even as the work's funeral attention to death is complemented by a parallel focus on youth and futurity.

The film's attention to the dialectical relationship between death and futurity is perhaps most evident near the end of the film, in the scene featuring Moon's "the living often like to pass judgment on the dead" voice-over. As Moon's voice-over is remarking, "I wonder how many people would understand how Ping, Sylvester, Susan, and I actually think?," the camera pans from a shot of his dead body slumped against Ping's grave, to a shot of the four children who had just discovered his body, and who are now flying a kite on a hill behind the grave. In this way, the four deceased young people referred to in Moon's remarks are visually replaced with another group of four younger children, who represent the future to which Moon's cohort no longer has direct access

As the radio broadcaster is quoting Mao's speech in this final sequence, the camera cuts from Moon's corpse back to a bare tree, in which the children's kite has got caught. The camera gazes up at the kite, which is decorated with a pair of large eyes that appear to be staring back down at the camera. The camera then rises up into the sky even as its gaze remains continually fixed on the kite in the tree below. In a curious optical illusion, even as the camera moves fluidly from a position directly below the kite to one high above it, the kite's two "eyes" appear to remain continually fixed on the camera throughout the entire shot—as if the kite were staring back at the camera even as the camera keeps its own virtual eye fixed on the kite.

The persistence of the kite's virtual gaze in this final shot contrasts with a set of sharp shifts in atmospheric lighting of the overall sequence. While the sequence begins with overcast skies, the shot of the three youngsters overlooking the cemetery has a much brighter lighting, while the final shot of the kite in the tree the sky is again overcast. Chan notes in an interview that his original plan was to have this final sequence conclude with an embrace between Moon and Ping, but he subsequently decided the embrace wasn't working the way he wanted and therefore went back later for some pickup shots. During the reshoot, however, the atmospheric lighting was significantly different from what it had been for the original shoot—though given the inherent logistical difficulties in setting up the shot in the first place, Chang decided to proceed with the reshoot anyway. When the resulting footage was spliced back onto the original sequence, the result featured erratic shifts in lighting—though Chan contends that since the viewer remains fixed on the two protagonists, the shifts in lighting are not really noticeable. In explaining his decision to proceed with the reshoot despite the change in lighting, Chan first notes that Hong Kong commercial film permits freedoms that normally wouldn't be allowed in an art film, while an "independent film," like *Made in Hong Kong*, allows the director even greater flexibility than he would have otherwise.

The persistence of the kite's virtual gaze in this final scene is contrasted not only with the final scene's shifts in lighting, but also with the rapid shifts in perspective featured in a 75-second montage sequence immediately following the voice-over reading of the Susan's, Ping's, and Moon's collective suicide note. From a close-up of the note itself, the camera cuts to a shot of a loose-leaf wall calendar blowing in the wind, then to a shot of Moon's white underwear hanging on a clothes line by the window. We then see some unidentified men running down the dark hallway of Moon's apartment complex, followed by a panning shot of a Chinese-language newspaper. Another shot of men running down a hallway is followed by a glimpse of the white liquid seeping out over pavement (as in Moon's dreams), followed by a shot of Moon's bedroom. A telephone is heard ringing offscreen, and the camera cuts to a shot of rivulets of red blood on pavement. Another shot of Moon's room is followed by another shot of rivulets of red blood streaming interminably across the pavement. There is a quick shot of Moon's face, wearing his signature sunglasses with yellow lenses, but this appears so briefly that it is almost imperceptible. The remainder of the sequence continues in this way, alternating between shots of the hallway, of Moon's bedroom and apartment, and of Moon's erotic dreams of Susan's suicide. There are voices offscreen of the police trying to page "account 818"—urging Moon to report to the station, and assuring him that they will offer him protection "regardless of what happened." Finally, the sequence concludes with a close-up of Moon's beeper, still clipped to his hip, which is beeping even as Moon's dead body is slumped against Ping's grave.

In this montage sequence, the pages of the wall calendar are blowing in such a way that the viewer can only make out the day, not the month and year. In the subsequent shot of the newspaper, however, viewers can make out the headline, which reads (in Chinese), "The Moment When Hong Kong Made History" (香港創造歷史的一刻). This headline clearly refers to the Handover, and features a verb, *chuangzao* (創造), that is closely related to the verb *zhizao* (製造) found in the film's own title. Just as the newspaper headline suggests that Hong Kong is not a passive pawn of historical forces beyond its control but rather is an agent actively "making history" in its own right, the Chinese title of Chan's could be parsed as either a passive construction ("made in Hong Kong"), as a sentence fragment featuring an active verb ("Hong Kong makes . . . "), or even as referring to "the making of Hong Kong" (with an elided "*de*," meaning "of"). Each of these three parsings of the film's Chinese title reflects a different dimension of Chan's film. As the newspaper headline at the end of the work suggests, *Made in Hong Kong* is very much concerned with this historical juncture in which Hong Kong finds itself in a position to "make history," and in this way remake itself. At the same time, the film clearly emphasizes its own origins and the fact that it is quite literally "made in Hong Kong." In other words, not only is Chan's film itself "made in Hong Kong," it simultaneously describes the making (or unmaking) of Hong Kong itself, through its thematization of the Hong Kong Handover and its implications for the former colony.

More generally, the Hong Kong Handover not only marks the film's historical backdrop and thematically anchors the film trilogy that began with *Made in Hong Kong*, it also offers a conceptual model for the structure of the work itself. That is to say, the Handover involves of process of conjoining two geopolitical entities such that they are viewed as one, just as cinema involves a process of stringing discrete filmic images together such they appear directly connected. The montage sequence at the end of the work, however, juxtaposes a series of short shots that cannot be logically linked together, and in fact do not even belong to the same representational plane—as it alternates back and forth between images of Moon's apartment and the hallway of his building, with other images taken directly from Moon's dreams, and even a couple of fleeting shots of Moon himself. Just as the montage sequence underscores the representational gaps that persist between its component shots, and the director's interweaving of leftover film stock from different sources points to that subtle differences in the appearance and text of the different film stock, similarly the work's dramatic exploration of the consequences of the Hong Kong's assignment to mainland control tacitly draws attention to the fissures that will continue to intervene between the two sociopolitical units.

PUBLIC TOILET

If *Made in Hong* marks the beginning of Fruit Chan's Handover trilogy, *Public Toilet* marks the end of the (truncated) trilogy that immediately followed it. After completing the second and third films in the Handover trilogy, *The Longest Summer* (去年煙花特別多; 1998) and *Little Cheung* (細路祥;

1999), Chan released *Durian, Durian* (榴槤飄飄; 2000), followed by *Hollywood Hong Kong* (香港有個荷里活; 2001). While the latter two works both prominently feature sex workers from Mainland China, and mark the beginning of Chan's so-called "prostitute trilogy," the "trilogy" was left incomplete with *Hollywood Hong Kong*. Instead, Chan's next film was *Public Toilet*, which took his oeuvre in a rather different direction.

Coproduced by the Korean production company Digital Nega and Fruit Chan's own production company Nicetop Independent Ltd. (the latter had produced all of Chan's films since *Made in Hong Kong*), *Public Toilet* features an international cast with actors from Hong Kong, South Korea, Japan, China, and Italy, and is shot in Beijing, Hong Kong, New York, South Korea, and India. The protagonist is young Beijing man by the name of Dong Dong (Chinese-Japanese actor Tsuyoshi Abe), who is born in a public outhouse. Dong Dong's mother had lost consciousness immediately following the birth and was taken away in an ambulance, after which an older woman had discovered the newborn in the cesspool beneath the bathhouse. The older woman then raised the infant as her own, and he comes to be known as the "god of toilets," in recognition of the peculiar circumstances of his birth. The film's main narrative plane unfolds eighteen years later, with Dong Dong and a group of friends hanging out in and around Beijing's outhouses. Several of these friends are foreigners who were born and grew up in Beijing, and they share with Dong Dong a sense of displacement resulting from the contingencies of their birth and a sense of indeterminancy with respect to their origins.

The main body of *Public Toilet* features several partially intertwined plotlines, including Dong Dong's travels to South Korea and the United States, as he attempts to find a cure for the woman who raised him, who is now dying; a couple of Hong Kong-born South Asian men who take their ailing father back to India to bathe in the Ganges; a Korean fisherman who finds a boneless mermaid who is sick from having ingested pollutants; and a young Hong Kong man who is in New York carrying out what he hopes will be his final assignment before he "retires." A focus on toilets and excrement is the primary unifying element in each of these disparate plotlines. For instance, the Hong Kong man's New York hit job goes tragically awry, and he ends up getting gunned down in a public bathroom, the Korean man finds the sick mermaid in a freestanding outdoor toilet near the ocean shore, and the film emphasizes that the Ganges River itself, though reputed to have magical healing properties, is in fact so full of sewage that it is effectively a public toilet in its own right.

A raw and jarring film, *Public Toilet* was also Chan's first full-length DV work. Not only does the digital technology permit Chan to film in multiple sites around the world on a relatively modest budget, it also yields a visual crispness that reinforces the work's quasi-documentary feel. More generally, to the extent that Chan's use of film (which generates an illusion of realistic movement through a rapid juxtaposition of discrete images) in *Made in Hong Kong* offers a useful conceptual model for assessing film's political significance (which is to say, its focus on the residual fissures generated by the forcible conjunction of Hong Kong and the PRC), his corresponding deployment of digital technologies in *Public Toilet* offers a similarly suggestive metaphor for the latter work's fascination with transnational connections. That is so say, just as DV is the product of a fundamentally mediated relationship between the external referent and its visual representation, the relationship between Hong Kong and the different geographical regions featured in *Public Toilet* is radically attenuated.

In this respect, the scatological comment of *Public Toilet* offers a suggestive metaphor for approaching both the work's cinematic medium. Just as excrement is the product of an elaborate digestive process that thoroughly transforms food's chemical structure, a digital image is similarly the product of a complex meditational process that leaves the image much further removed from the original visual stimuli (unlike film, in which the relationship between visual stimuli and filmic image is basically indexical). The sequence in *Public Toilet* that most explicitly foregrounds the film's digital medium occurs in the latter portion of the work, when Dong Dong is in New York and meets the Hong Kong hit man (played by Sam Lee). The hit man gives Dong Dong a handheld video camera, and asks him to promise to film everything that happens next. Then, the hit man proceeds down the street, enters a building, then reemerges moments later—having been shot and mortally wounded. Still holding the video camera, Dong Dong then rushes into the same building the hit man had entered, goes into a public restroom, and grabs a second video camera that he finds mounted on the wall, and runs out with it. After he is safely away from the scene of the killing, he finally stops to view the footage recorded on the restroom camera.

Not only does this New York sequence revolve around an exchange of video cameras, significant portions of it are presented as though they were being recorded by those same cameras in real time. For instance, the shot of the hit man proceeding down the street is presented on screen just as Dong Dong would have seen it in the camera's viewfinder—complete with icons indicating that the machine is recording, and so forth. Later, when Dong Dong is viewing the footage that had been recorded by the camera left in the public restroom, the shot is presented as Dong Dong would have viewed it on the camera's built-in display—complete with miniature icons indicating the camera's battery status, and so forth. After Dong Dong finishes watching the video, he immediately throws the camera over the railing of the bridge on which he is standing, and we see the camera's trajectory through the air and into the water as though recorded by the camera itself.

The second video camera (which was presumably placed in the restroom beforehand by the hit man himself), reveals that after the hit man ran entered the restroom, he tried to retrieve a handgun he had hidden inside the tank of one of the toilets. It turns out, however, that a homeless man (who is still in the restroom when the hit man enters) had already found the gun and removed it, as a result of which the hit man is left defenseless and is fatally shot by the same men he had been sent to kill. This sequence in the restroom echoes an earlier subplot in the film, in which a boy suffering from leukemia finds a handgun hidden in the toilet tank in a hospital bathroom, and also directly mirrors a similar sequence from *Made in Hong Kong*, in which Moon retrieves his handgun from a toilet tank before attempting to carry out his assassination of the two businessmen at the end of the film. In this way, the circulation of the figure of the handgun within *Public Toilet* mirrors the way in which the handgun functions as a thematic hinge between Fruit Chan's 2002 film and his earlier *Made in Hong Kong*.

Just as *Made in Hong Kong* unfolds under the shadow of Susan's suicide and Ping's terminal illness, *Public Toilet* similarly follows the intersecting trajectories of several characters who are each trying to find a cure for a loved ones; and just as *Made in Hong Kong* concludes with a shot of the cemetery where Ping is buried and Moon has committed suicide, *Public Toilet* also ends on a similar note. In particular, before the final credits a message appears on screen specifying that the film is dedicated in memory of the actor Pietro Diletti, who played the part of a young Italian man who was born and grew up in Beijing, but who had decided to return to Italy for college because he wants to "understand Italian history." Diletti's character appears several times in the initial portion of the work, but the final memorial note points out that he died in a car accident in Italy in 2000—meaning that Fruit Chan must have started working on *Public Toilet* even before finishing the first installment in his "prostitute trilogy," *Little Cheung* (which had its debut in April of 2000). As a result, it is almost as if Diletti's contribution to the film lies not so much in his actual presence, but rather in the way that his untimely demise anticipated the work's more general concern with death and mortality, together with the way in which his decision to return to Italy to recover his roots informs the work's broader fascination of themes of geocultural displacement and questions of identity.

NOTES

1. "*Daoyan Chen Guo fangtan: Haokan de dianying dou shi hao dianying*" [Interview with Fruit Chan: Films that are enjoyable to watch are all good films], *Jinghua shibao*, January 14, 2002. http://www.china.com.cn/chinese/CU-c/96954.htm. Here and throughout, translation is mine, unless otherwise noted.

2. Michael Berry, *Speaking in Images: Interviews with Contemporary Chinese Filmmakers* (New York: Columbia University Press, 2005), 466.

3. Esther Cheung, *Fruit Chan's Made in Hong Kong* (Hong Kong: University of Hong Kong Press, 2009), 5.

4. For a detailed discussion of the *Young and Restless* series in relationship to *Made in Hong Kong*, see Esther Cheung, *Fruit Chan's Made in Hong Kong*, 56 ff.

5. Esther Cheung, *Fruit Chan's Made in Hong Kong*, 5–6.

6. Susanna T, "No Future, No Future! Fruit Chan Speaks about *Made in Hong Kong*." In *Hong Kong Panorama 97–98* (Hong Kong: Provisional Urban Council, 1998), 54–57. This claim, however, should be taken with a grain of salt, since some of the voice-over occurs in places where no significant acting is called for (as in the opening scene, where all we see is a medium shot of Moon playing basketball with his friends).

7. Esther Cheung, *Fruit Chan's Made in Hong Kong*, 121–122. A similar argument is often made that Taiwan director Hou Hsiao-hsien developed his distinctive reliance on medium- and long-shots in order to deemphasize the limited training of his nonprofessional actors.

8. Here and throughout, quotes from the film are adopted from the work's English-language subtitles. In many cases, the translation has been revised.

9. Mao Zedong, "American Imperialism is a Paper Tiger," November 17. 1947.

Quoted in "Chinese-Soviet," March 7, 2013. http://www.360doc.com/content/13/0307/21/7915662_270057022.shtml

10. See Li Jian, "*Yangshi jiemi ZhongSu enyuan chedi cuihui Mao Zedong*" [CCTV deciphers the resentments between China and the Soviet Union; Completely destroys Mao Zedong], *Xin Tangren dianshitai* [New Tang Dynasty Television], March 2, 1013. http://www.ntdtv.com/xtr/gb/2013/03/02/atext855781.html. Article reprinted in English in Ariel Tian, "Mao's 'Nuclear Mass Extinction Speech' Aired on Chinese TV," *Epoch Times*, March 5, 2013. http://www.theepochtimes.com/n3/4758-maos-nuclear-mass-extinction-speech-aired-on-chinese-tv/

BIBLIOGRAPHY

Berry, Michael. *Speaking in Images: Interviews with Contemporary Chinese Filmmakers*. New York: Columbia University Press, 2005.

Cheung, Esther. *Fruit Chan's Made in Hong Kong*. Hong Kong: University of Hong Kong Press, 2009.

"*Daoyan Chen Guo fangtan: Haokan de dianying dou shi hao dianying*" [Interview with Fruit Chan: Films that are enjoyable to watch are all good films], *Jinghua shibao*, January 14, 2002.

Li Jian, "*Yangshi jiemi ZhongSu enyuan; chedi cuihui Mao Zedong*" [CCTV deciphers the resentments between China and the Soviet Union; Completely destroys Mao Zedong], *Xin Tangren dianshitai* [New Tang Dynasty Television], March 2, 1013. http://www.ntdtv.com/xtr/gb/2013/03/02/atext855781.html. Article reprinted in English in Ariel Tian, "Mao's 'Nuclear Mass Extinction Speech' Aired on

Chinese TV," *Epoch Times*, March 5, 2013. http://www.theepochtimes.com/n3/4758-maos-nuclear-mass-extinction-speech-aired-on-chinese-tv/.

Mao Zedong. "American Imperialism is a Paper Tiger," November 17, 1947. Quoted in "Chinese-Soviet," March 7, 2013. http://www.360doc.com/content/13/0307/21/7915662_270057022.shtml

T, Susanna. "No Future, No Future! Fruit Chan Speaks about *Made in Hong Kong*." In *Hong Kong Panorama 97-98*. Hong Kong: Provisional Urban Council, 1998. 54–57.

Chapter 8

The Transnational Gothic Visions of Luis Buñuel and Gabriel Figueroa

Ceri Higgins

The complexity of the transnational collaboration between Spanish director Luis Buñuel and his preferred[1] director of photography, Mexican Gabriel Figueroa, was expressed through a darkly, Gothic vision. Creating a highly constructed visual paradigm, they probed and revealed social and psychological contradictions and conflicts in narratives and characters to make an incisive critique of the sociopolitical contexts within which they made their films. Their creative partnership established a visual and technical model, that was to influence subsequent filmmakers in Mexico, to the present day.

This chapter will examine the Buñuel/Figueroa creation of a transnational, Gothic visual narrative and its subsequent development by filmmakers, focusing primarily on the work of Guillermo del Toro. As Williams succinctly summarizes, "Gothic is a discourse that shows the cracks in the system that constitutes consciousness, reality" (Williams: 66). It is significant the Buñuel/Figueroa collaboration drew on the Gothic, an intrinsically transnational genre, to construct their vision.[2]

From its roots in eighteenth-century Britain, Gothic has developed into a potent creative force in literary and visual culture; from architecture, design, fine art, film, games and explored in virtual world applications.[3] According to Enrique Ajuria Ibarra, "Gothic has been intrinsically characterized by its capacity to move and travel and its ability to fascinate and terrify a wide transnational audience" (Ajuria Ibarra: 6).[4]

Discussion of the Gothic in cinema has tended to concentrate on narrative themes in the films, drawn from and relating them to gothic literature; giving an emphasis on the uncanny, supernatural and fantastic, and centered on loss, grief, madness, creation, death, and "otherness." Narrative contradiction and characters are analyzed together with the trope of duality, multiple personalities and doppelgangers. However, as Jonathan Risner proposes in his essay, "Filmic Spaces in Contemporary Latin American Gothic Horror Cinema," the cinematic gothic is essentially a set of "visual codes." These repeated motifs signify the formal elements of Gothic, through cinematography: mise-en-scene, lighting and composition (Risner, 39). Risner provides a compelling discussion of space in three contemporary Latin American films, *Visitante del invierno* (2008, Esquenazi), *La casa del fin del los tiempos* (2013, Hidalgo) and *La casa muda* (2010, Hernandez), drawing on, as a methodology, Bachelard's "the systematic psychological study of the sites of our intimate lives," his "topoanalysis" taken from *Poetics of Space* (1994). The main points of Risner's discussion of contemporary cinema also resonate with the much earlier work of Buñuel and Figueroa. What is notable is that the twenty-first century Latin American films Risner analyses, with their use of digital cinematographic techniques, continue to use and build upon the visual language of the Gothic as established in early European cinema. A language that through transnational cultural migration, caused by political and economic changes in Europe in the early twentieth century, crossed the Atlantic, and via Hollywood synchronized with local cultures and traditions to develop a Latin American Gothic. The personal trajectories of Buñuel and Figueroa made them integral to this transnational migration of filmmakers, cinematic genres and practices, including the Gothic and consequently places them firmly at the center of a transnational and Mexican cinematic gothic.

Throughout their lives Buñuel and Figueroa, had crossed national, political and cultural borders and in their work together, they contravened social, psychological and sexual barriers. Through their films they argued that resistance to transgression, the keeping to "safe familiar territory" could become a prison and lead to internal confusion, repression, madness and death. The constructed spaces the characters occupy, and the light that illuminates them and the composition and framing that externalize these conflicts, make physical the interior psychological spaces the characters inhabit and that they attempt and/or fail to transgress. The failure to transgress, in the Buñuel/ Figueroa frame, is a critique of social breakdown:

> The discourse of degeneration also partakes of a certain Gothic fluidity [. . .] And what links most of the debates is [. . .] the drive to define and categorise the features of a culture in crisis, to determine the exact nature of the agents of dissolution and decline. (Byron, 2000: 133)

However, unlike the classic Gothic text where there might be a "desire to redefine and fix a 'norm', to re-establish the boundaries that the threatening other seems to disrupt and destabilise" (Byron, ibid), for Buñuel and Figueroa there is a desire to destroy boundaries altogether.

The act of crossing boundaries and to go beyond accepted norms and the Gothic tropes of otherness and exclusion, resonate deeply with Buñuel and Figueroa's personal experience. In different ways, both filmmakers were exiles. Buñuel spent most of his adult life outside of his native Spain and Figueroa's social hybridity accentuated the displacement integral to the experience of being Mexican. I connect this sense of "otherness" to Julia Kristeva's ideas in *Strangers to Ourselves* and Edward Said's *Reflections on Exile,* and would suggest that an experience of living on the periphery relates directly to a Gothic sensibility.

Kristeva sees the process of displacement as the experience of the "stranger"whom she argues is as much internal as external. She/he is a sum of the social and psychological inconsistencies, integral to all of us, that we can refuse to accept or to which we can submit, but which we can never ignore. Continual transformation is the central dynamic of "the stranger" who inhabits a constantly changing, transitory space, on an infinite journey. The contradictions manifest in successive Mexican political regimes and their ideologies are fundamental to the way the nation has developed and echo Kristeva's notion of the continual transfiguration of internal/external space. As a Mexican, Figueroa may be seen like Kristeva's outsider, the stranger who is a sum of internal contradictions brought about by his social context, yet who is also in a state of constant transformation. Indeed, Figueroa stated that his transfiguration of reality through the lens of a camera led to a transformation in himself.[5]

In his compelling and moving set of essays, *Reflections on Exile*, Edward Said observes that exile is "the unhealable rift forced between a human being and a native place, between self and its true home" (Said, 2001: 173). His statement brings together the notion of external exile from place with the idea of internal exile from self. Significantly, Said goes on to describe the essential link between exile and nationalism as being "like Hegel's dialectic of servant and master, opposites informing and constituting each other" and nationalism, he argues, is a result of alienation (2000: 176). The innate schism in the idea of estrangement evokes the fissures and rifts embodied by Kristeva's stranger and that corresponds to an understanding of the complexities of Mexican nationalism, which I have discussed elsewhere (Higgins, 2008).

Following on from Said's and Kristeva's notions of rift, alienation and displacement, I suggest that exile is innate to the Mexican experience. Mexicans encounter displacement from their own space, time and culture in the same way as émigrés. Said's rift between "self and a true home" resonates with the constant and consistent encounter with the "other" in Mexico between the internal self (physiological, psychological and emotional as *mestizos*) and the external home (the everyday position in a post-colonial space) and results in a constant, simultaneous exile and homecoming. Being a stranger in one's own land becomes a dialogic existence that Figueroa externalizes through the lens.

Contradiction is central to Buñuel's work and exile, "an insoluble contradiction" is a "main creative force" in his films (Fuentes, 2004: 159). Significantly, the recent reappropriation of Buñuel in a Spain that has ignored the question of the director's status as an exile, an important absence when one considers that Buñuel was resident in Spain only until his twenties and, indeed, in his forties became a Mexican citizen (Fuentes, 2004: 159). Moreover, of the thirty-two films he directed, twenty were Mexican. Consequently, the major part of Buñuel's creative output as a director was developed and made in his adopted country (Pérez Turrent, 1993: 62). I propose that exile was also a compulsion in the imagination and creativity of Figueroa and that this mutual expression of exile, both internal and external, drew Buñuel to collaborate with Figueroa more than any other director of photography.

While the notion of exile has, in general, been ignored in relation to Buñuel (the notable exception being Fuentes's analysis), consideration of Figueroa as displaced might appear downright incongruous. Although not obviously an exile, Figueroa's movement between classes, his adoption of oppositional politics and his social hybridity made him an internal émigré. Victor Fuentes defines exile as "being in one place, but to have one's imagination focused elsewhere" (2004: 159). The ambiguity of Figueroa's social, political and professional position within Mexico, together with his complex situation in relation to Hollywood and the United States, was the foundation upon which he created images of a Mexico that was "elsewhere." As I have already suggested previously, Figueroa's images were at once present in the imagination of the Mexican audiences, yet topographically absent (Higgins, 2004: 216–217). Indeed, it is a paradox that links Figueroa with Buñuel as transnational, "extraterritorial" filmmakers and chimes with Fuentes's notion of the "insoluble contradiction" of exile (Fuentes, 2004: 159).

Close visual and narrative analysis of *Los olvidados* (1950), *Él* (1953) and *El ángel exterminador* (1962) reveals the visual language Buñuel and Figueroa constructed to give a subversively Gothic insight into the internal workings and demise of the individual and society. I will go on to propose that their style resonated with succeeding directors and cinematographers to be a central influence in the cinematic expressions of a transnational Gothic.

Since Luis Buñuel's debut as a filmmaker in 1928, a large body of literature has emerged, conferences organized and, in more recent years, websites established, that analyze the work of this Spanish surrealist from a range of psychological, political and social perspectives. However, as scholar Ernesto Acevedo-Muñoz has observed, most of the critical attention concentrates on what is perceived as two "peaks" of Buñuel's career: his surrealist trilogy of the late 1920s to early 1930s (*Un chien andalou* (1928), *L'Âge d'Or* (1930) and *Las Hurdes* (1932)), and the Spanish and French films made during the 1960s and 1970s (Acevedo-Muñoz, 2003: 2). The work which has had less attention is paradoxically within his most productive period – the twenty films he made during the late 1940s through to the early 1960s in Mexico where he had taken up permanent residence and indeed had become a Mexican citizen,

In writing about Buñuel's most well-known film of this period *Los Olvidados* (1950), which won a best director award for Buñuel in the 1951 Cannes Film Festival, Peter Evans states, "*Los olvidados* moves beyond the prose of documentary and into the Mexican Gothic, transforming dross into metaphor, the ordinary into the fantastic, the known into the unknown and disturbing" (1995: 78). Evans employs a Freudian psychoanalytic paradigm to analyze the director's work and goes on to make a connection between gothic tradition in *Los olvidados* and Freud's writings on the uncanny.

Rather than add to the by now considerable body of psychoanalytic analyses of Buñuel's work, I analyze the gothic tradition evident in his films from a transnational and sociopolitical standpoint.

> In Buñuel's movies the seen and the ordinarily unseen inhabit the same film space; he pictures the picturable, and strongly alludes to what cannot be pictured.
>
> (Wood, 1993: 44)

Michael Wood's perceptive observation of Buñuel's images leads one to consider the construction of space in his films. In the following analysis of *Los olvidados*, I investigate how Figueroa's use of light and shadow together with contrasts in space and composition, communicate complex themes that interweave with the displacement of the mother figure. What strikes the viewer immediately about *Los olvidados* are the visual contrasts. The oppositions of darkness and light, the vast empty wastelands and cramped interiors, day and night. Throughout the film, these oppositions are set against each other to build a visual dialectic that reflects the contradictions and conflicts within the characters. The exteriors are vast, lit by natural, hard, flat light of an urban wilderness. Construction is abandoned and only the scaffold skeletons of modernity are left, the anonymity and timelessness of the urban space captured by Figueroa's choice to shoot at midday when the harsh, shadowless sunlight flattens perspective.

In contrast to the exterior scenes in the interiors, Figueroa emphasizes the claustrophobic, crowded and cramped conditions. Key lights, focused on small areas, confine space and the textured shadows that fall away delineate the edges of the interior spaces. Further, many compositions take in the ceiling and walls and wall-to-wall beds dominate every room, their iron frames throw shadows that entrap characters in prison-like bars of light and dark. The interior space is the site for the heightened emotional scenes. The shadowy barn, with its half-hidden animals is a primal space that threatens rather than comforts. It is a place in which the characters hide or seek refuge and it is the site of the ultimate struggle between Jaibo and Pedro, that culminates in Pedro's death. Yet, despite the cramped framing and compositions of the interiors and the use of group mid-shots and long-shots, Figueroa's lighting also separates characters to emphasize their isolation. He creates a paradox of solitude and claustrophobia with a key light focused on the main area of action and low-intensity fill lights which leave the dark corners in shadow and backgrounds beyond the main action in darkness.

One of the best examples of how Figueroa works with the narrative contradictions in the film is the scene between Jaibo and Pedro in the cutler's shop. The main oppositions are within the characters themselves. The boys each have light and dark elements to their personalities which are constantly in flux. It is an ebb and flow which creates in the viewer a visceral awareness of the gray areas between, and within, the two to generate in a compassionate detachment. Like two planets in orbit, each one moves constantly from near/light side to far/dark side within themselves (internally) and to each other (externally). Each boy is a double (doppelganger) of the other and the highly theatrical space that Figueroa constructs in the cutler's workshop is a visual manifestation of the gothic fluctuation between the internal and external, insider and outsider, light and shadow.

With the use of one backlight as the key source in the scene, Figueroa layers the shop with light, directing the viewer's eye from the darkened foreground over the lit table and the glittering knives to the straight, dark angles of the bellows and Pedro silhouetted, hazy in the wood smoke, against the backlight. Figueroa isolates the interior from the exterior. The high contrast created by a strong backlight as the main source through the exterior door makes silhouettes which together with the glow of the fire and the smoky atmosphere constructs a Niebelungen-like cave, underscored on the soundtrack with the rhythmic clanging of a hammer on white-hot steel. Jaibo enters, silhouetted against the door frame, a diabolic, faceless figure, barely recognizable through the smoke. Throughout the scene, the boys move between areas of light and shadow, from silhouette to distinct form, in a visually gothic struggle between good and evil, circumstance and opportunity. Figueroa frames the boys throughout nearly all of the 2-shots with one of them against a sharply focused source light and the other against a defined area of dark.

The viewer never sees the boys together in Pedro's home, yet they both have pivotal scenes with the mother in the cramped communal dwelling. The mother aggressively rejects Pedro, yet passively and then seductively accepts Jaibo. In these scenes each boy is lit and framed to reflect the complex working through of the son-mother/male-female relationship. With Pedro, this complex turns on an ambiguous conflict with the mother and with Jaibo it is the development from a surrogate son to a lover. Figueroa lights Pedro as luminously innocent, with a soft three-point lighting setup and use of a thin gauze over the lens, through to framing him in a hard directional key light, to create contrast and shadow, to reflect Pedro's aggression and confusion. On the other hand, in the same physical space, Jaibo continuously moves back and forth between strong, key light and dark shadow. In the subsequent seduction scene the lighting and focal depth soften both the mother and Jaibo, to isolate the pair from their surroundings and to draw them together in the complicity of seduction.

In Mexico, the mother is perceived as an icon, the foundation of national creation myths, both the mother of God in Coatlicue and the betrayed Mother in Malintzin. *Los olvidados* is essentially about the lost or absent mother. Pedro is rejected by his mother, Ojitos has only a father who is also absent and he is left to suckle a goat as his substitute nurturer. Meche's mother is an invalid and confined to bed and Jaibo's lack of mother leaves him with only a fantasy of her, that of the *Virgen*. I argue that Figueroa's visual representation of Jaibo and Pedro expresses aspects of poverty, social injustice and the deep contradictions in the foundation myth of the mother and its effects on Mexican national identity. The social reality shown in the film challenges the myth of the mother. For this reason, Buñuel cannot be seen, to ally himself with social realism, because in this film and, indeed, all of his work, myth and its gothic expression is equally pervasive in and influential on reality as social conditions and circumstances. Indeed, "*Los olvidados* is something more than a realist film, since in it dreams, desires, horror, delirium and chance, the dark side of life, also play a part" (Paz: 1961).

In *Los olvidados* Buñuel and Figueroa attack the gut of Mexico to present the viewer with a world in which mothers are absent, or worse, reject their offspring. In showing a society where the mother-virgin, becomes the mother-lover, he presents an Oedipal society. Buñuel's mother-virgin-lovers are not the martyrs one finds in Fernández films which comfortably support the national *complejo* of veneration and victimization. The tough presentation of the Mexican mother in Buñuel's films suggests a route to subversion, which if followed would lead to a true revolution in the Mexican psyche. This is why, in exposing the imperceptible chasm between Mexican myth and Mexican social reality, in *Los olvidados,* Buñuel was so heavily condemned in Mexico.[6]

Figueroa establishes the binary paradoxes of the film visually, working with the director to present the viewer with two alternative resolutions in Jaibo and Pedro, each of which ends in darkness and death. Rather than simply establish the two characters as narrative and visual opposites, Buñuel and Figueroa work together to communicate the complex internal conflicts in the individual that result from poverty and social injustice. The visual presentation of *Los olvidados* may superficially resemble the documentary aesthetic of the Italian neorealist style but under closer, more detailed analysis of the cinematography it is clear that the visual style plays with the cinematic conventions melodrama and German expressionism, which Figueroa and Buñuel develop into a Gothic visual language.

Consequently, through lighting and composition, Figueroa makes explicit profound levels of meaning that constitute Buñuel's worldview.

Darkness and death, so fundamental in the content, form and visual language of *Los olvidados* is central to the gothic tradition. In *El angel exterminador* and *Él*, Buñuel and Figueroa develop the gothic in relation not to the underclass, but to the tradition's established social environment, the ubiquitous mansion of the bourgeoisie and the ruling elite.

The Bourgeoisie in *Él* and *El ángel exterminador*

> I don't believe anyone is morally determined forever because he was born in such-and-such a social class. Being born bourgeois doesn't condemn anyone to think or behave like a bourgeois for his entire life. Co-existence changes one's manner of being. [. . .] If you and I were forcibly locked up together, [. . .] we would almost certainly end up hating each other. (Luis Buñuel)[7]

In *El ángel exterminador* (1961) a group of dinner guests are inexplicably unable to leave the music room of their host's house. Over the days and weeks they are trapped, the neurotic turmoil of their internal selves erupts and destroys the well-mannered superficiality of their external behavior. The eruption of psychosis is the main theme of *Él*, in which the protagonist, the upstanding pillar of the church Francisco Galván de Montemayor (Arturo de Cordóva), crumbles into a paranoid delirium during the course of which he loses his hold on reality and attempts to circumcise his wife.

Both films examine and dissect the capitalist class, the owners of society's means of production and bulk of its wealth. The term bourgeois carries with it an inherent sense of materialist values and conventional attitudes. Its root dates back to the French sixteenth-century, late-Latin word *burgus*, meaning castle or fortified house. Buñuel's incisive definition of the bourgeoisie takes the home/*burgus* as fundamental to his depiction of a ruling class that jealously guards the privilege of its social space, the material expression of which is the house, the mansion or the stately home. Buñuel reveals that in its protection of wealth and social advantage, the bourgeoisie imprisons itself literally within the walls of the home while it metaphorically incarcerates itself within class ideology.

The *burgus* is central to Buñuel's narratives and is the site where characters unravel and degenerate. In *El ángel exterminador*, dinner party guests are inexplicably trapped in the music room of a mansion and in *Él*, the overwhelming presence of an eccentric house and Francisco's frenzied pursuit of a hopeless legal case to regain lost family properties traps him into a cycle of repression, violence and potential castration. Entrapment in the confines of the *burgus* forces the characters to express the deep values and morals that lurk underneath the niceties of acceptable behavior. In so doing, Buñuel's characters either slip into paranoia and psychosis or die. Consequently, the construction of space is central to Buñuel's vision of how the bourgeoisie functions and is the key to its potential collapse. As cinematographer, Figueroa constructs the filmic space that constitutes the *burgus*. In both films Figueroa visually links the characters' relationship with exterior, physical space to their interior emotional and psychological state. Therefore, a close analysis of the way in which the cinematographer manipulates space in both *Él* and *El ángel exterminador* illustrates the Buñuelian definition of the bourgeoisie and consequently heightens the subversive impact of both films.

These films, made ten years apart, demonstrate how Buñuel, together with his long-term Figueroa, developed a gothic visual style in relation to not only the urban underclass and modernity as in *Los Olvidados*, but to the tradition's more established social environment, the ubiquitous mansions of the bourgeoisie and the ruling elite.

Gothic convention is intrinsic to the construction of the bourgeois narratives and visual spaces in both films. Indeed, there is a direct reference to the Gothic in the opening scenes of *El ángel exterminador*. The host, Edmundo Nóbile, proposes a toast to Sylvia, one of the guests, for her performance in Donizetti's opera *Lucia di Lammermoor*. The opera is an adaptation of Walter Scott's gothic novel, *The Bride of Lammermoor* which comments on the demise of the ruling class brought about by extreme changes in the main characters of the book and in external events.

Scott's nineteenth-century novel, in keeping with other gothic narratives of the period, is a critical examination of the conflicts between the *nouveau riche* nobility, the traditional, aristocratic landowners and the developing bourgeoisie, themes that also resonate throughout Buñuel's work.

El ángel exterminador and *Él* reverberate with gothic motifs. The primal space of the castle/mansion and the key elements of the curse, darkness and omens, internal conflict, madness and physical and psychic changes of state are present in both films. Further, the key gothic tropes of duality, entrapment and the darker, inner self that overcomes the outer "civilised" self are central to the development of the films' main characters. In keeping with Buñuel's vision of the bourgeoisie, Figueroa developed a visual language drawn from the motifs and style of German expressionist cinema, which he learned from his mentor Gregg Toland. Toland (best known for his work with Orson Welles (*Citizen Kane*, 1941) William Wyler (*Wuthering Heights*, 1939) and John Ford (*The Grapes of Wrath*, 1940)) was a former apprentice of the renowned cinematographer Karl Freund who believed "the cameraman ought to create shadow. That is much more important than creating light" (Sears, 2003: 170). In his early films, *Der Januskopf* (*Doctor Jekyll and Mr Hyde*) (Murnau, 1920) and *Metropolis* (Lang, 1927), Freund developed what Fred Botting calls the "modernist Gothic" in film. This distinct visual style draws on the narratives and archetypes of the gothic novel to create an imagery of high contrast, heavy shadows, acute angles and distorted sets to express the dark, supernatural and political themes inherent in the gothic tradition (Botting, 1996: 165–168). In Hollywood Freund developed the "modernist Gothic" further and in so doing founded the signature style of the horror genre through the seminal films *Dracula* (Browning, 1931) and *Murders in the Rue Morgue* (Florey, 1932) and his directorial works *The Mummy* (1932) and *Mad Love* (The Hands of Orlac) (1935). Buñuel, together with the surrealist group in Paris, had a fascination for the gothic novel that fed into surrealism through symbolism. This, combined with the direct influence of Karl Freund and Toland on Figueroa, provoked an elaborate interplay of themes, textures and space in *Él* and *El ángel exterminador* that went on to further develop "modernist Gothic" imagery.

GOTHIC STYLE AND ALIENATION

In his personal life, Buñuel, crossed national, political and cultural borders and in his work he transgressed social, psychological and sexual barriers. Through his characters he argued that resistance to transgression, the keeping to "safe familiar territory" could become a prison and lead to internal confusion, repression, madness and death. In *El angel exterminador*, the characters' inability to leave a room (despite there being no physical barrier, not even a door), traps them into a descending spiral of deterioration to finally expose their true selves. The characters have an innate sense of entitlement. They are all in positions of power and the dinner guests are major figures in politics, the sciences and culture; their sense of power, ownership and dominance symbolized by the mansion, which significantly entraps them. The film opens with the servants, acting on a gut feeling of foreboding, leaving one by one, as the dinner guests arrive. Only the butler, Julio, remains, himself not a bourgeois, but a strong supporter of the status quo—a representative of hegemonic forces which enable the bourgeoisie to maintain its dominance.

When, the guests first realize they are unable to leave the room, the lighting becomes more expressionist and the social niceties break down. The entrance to the music room from the adjacent drawing room becomes a proscenium arch and the room beyond the stage on which the action unfolds. Later, during a gothic-like thunderstorm, Figueroa positions the camera in a wide-angle wide-shot from the drawing room to the music room. The foreground drawing room remains in darkness, lit only by the occasional flash of lightning.

In the background of the frame the viewer sees the music room like a brightly lit stage in the darkness. The guests move around as aimless as the sound of the piano. The sense of watching a piece of theater is once again evoked, but in the following shot the viewer is back on the stage with the players in a reverse angle of the wide shot. Figueroa also uses tracks to take the camera through the arch that divides the music and drawing rooms and as a result accentuates the viewer's position as both inside and outside the action. Further, the absence of conventional point-of-view shots and reverse angle shots in the film prevent the viewer from being drawn into an empathetic relationship with the characters and the tracking shots distance the viewer from the action to create a cinematographic *alienation effect*.

The view of the music room and the characters' place within it is constantly challenged. The gothic struggle between the inner and outer self is made visually manifest as the narrative unfolds, leading to a series of dramatic hallucinations and visions as the guests' inner thoughts and emotions are revealed. A variety of angles and changes in lens provide continual shifts in perspective. On one hand, long-shots and mid-shots on a wide-angle lens with lighting that provides expressionistic layers of light and dark to create depth and width to the space. On the other, there are mid-close ups and close-ups shot on a short lens that flatten depth of field, to isolate characters from the space around them and each other. This change of depth and perspective remains consistent throughout the film. The visual dialectic that builds up between confinement and space captures the characters' contradictory responses to their situation and each other. It visually expresses the fundamental gothic trope of constant vacillation from the inner to outer selves demonstrated by the guests dealings with each other and themselves, fluctuating from communication to secrecy, honesty to deceit, love to hate, decency to perversity, life to death. As the characters' darker, inner selves emerge the space becomes more expressionist with background often dropping into complete darkness and individuals lit in subtle pools of directional light. The bourgeois space finally becomes the gothic *burgus*.

THE *BURGUS* AS MADNESS IN *ÉL*

In *Él* (1951), Francisco's outward formality and need for order stands in direct contrast to the style of his mansion. Designed by his father, the house is inspired by Art Nouveau. There are no straight lines in the architecture, the graceful curves and sensuality of the style made popular during the Porfirian era, is exaggerated in *Él* to provide a subtle sense of spatial disequilibrium. In architectural terms the interior of the house lacks harmony and conventional proportion and the spaces are disconcertingly awry and claustrophobic.

The rich chaos of line contrasts dramatically with the rigid symmetry of the stairway that dominates many scenes in the film as a central element of the mise-en-scène and are a physical representation of Francisco's stasis within his erratic and unstructured psychosis, represented by the flowing curves of the Art Nouveau that surrounds the straight, heavy lines of the steps. Further, the stairway is where Francisco's mental state becomes most evident in the scene in which he paces to and fro across the confines of the stairs, banging the railings with a stick, in a fit of frustrated manic rage.

Significantly, the exterior of the mansion is only seen in the brief scenes of the garden and when Francisco spies on his wife, Gloria (Delia Garcés) from the high fortress like terrace. Like the mansion in *El ángel exterminador*, a high wall surrounds the house. Francisco's home is indeed the *burgus*, a luxurious lair that both protects and entraps.

However, in *Él*, more so than in *El ángel exterminador*, the interior represents the inner life of the owner. In *El ángel exterminador* the exterior becomes a site of spectacle as crowds gather to wonder at the plight and reasons for the guests entrapment. By contrast, in *Él* the viewer experiences Francisco's psychosis from within. Francisco is a paradigmatic gothic character. His outer self is a paragon of decency and an exemplar of bourgeois noblesse oblige. His inner self, that overcomes the controlled exterior, is dark and psychotic. The cause for the internal character to surface is the imminent loss of the *burgus* and the impending threat of modern values and ideology, represented by the introduction of the engineer Raúl Conde (Luis Beristaín), into Francisco's reactionary world.

Whereas in *El ángel exterminador* the *burgus* is a ubiquitous physical presence that entraps the characters within the music room, in *Él* the mansion is a manifestation of the larger, lost *burgus* that Francisco frantically tries to recuperate. As a bourgeois, his social, physical and psychological identity is intimately linked with the *burgus* and the increasing futility of his legal battle to regain family property lost in the revolution is integral to his decline into psychosis.

Francisco and Gloria are seen through a system of expressionist lighting, angles and framing. Gloria is lit with a soft backlight to create a halo-like effect. The diffuse key and fill light soften her face and give it a luminous quality. By contrast, a strong key lights Franscisco at an acute angle to the side of his face. The position of the light creates a hard shadow that bisects the face to give Francisco a literal light and dark side. The use of diffuse backlight on one character and strong key light on the other in a two-shot is significant as it directly challenges the convention that characters should be equally lit in shot. In the choice to shoot the scenes between Gloria and Francisco in this way, Buñuel and Figueroa demonstrate a clear commitment to build character and narrative tension through careful construction of a modernist gothic image. Further, in the Francisco-Gloria two-shots, shadows, architectural elements or furniture provide subtle barriers and divide the couple in frame. Francisco dominates the two shots with Gloria and combined with a consistent use of low-angle close-ups, Francisco's overpowering presence pervades the film.

As Gloria and Francisco's relationship develops, the light and dark, high angle and low angle camera positions and use of depth of field express the deterioration of the couple and the destructive dynamic between them caused by Francisco's psychotic behavior and which culminates in his visual and narrative breakdown in the cathedral bell tower. A wide-angle lens accentuates the acute low angle and the size of the bell. Significantly, this final destruction of Francisco's marriage and sanity takes place in this archetypal gothic location. In gothic tradition, churches and towers are spaces which simultaneously isolate and empower the characters and the inherent blasphemy of characters such as Count Dracula and Doctor Frankenstein are echoed in Francisco's psychotic declarations. The bell tower is also an extension of Francisco's *burgus*. He regards it as his natural home and his position as naturally equal to God. On his return to his mansion, Francisco paces the stairway, running and banging a stick back and forth over them. The use of exaggerated depth of field, the huge, bar-like shadows that run up the stairs and over the hallway and the straight symmetrical lines of light and dark obscure the erratic curves of the Art Nouveau interior. Francisco entrapment on the rigid lines of the steps and his inability to move away from them communicates his impotence and frustration.

In his visual construction of the mansion and other spaces that represent the *burgus* in the film, expressionist techniques demonstrate the psychosis and isolation of Francisco. Like the guests in *El ángel exterminador*, Francisco is marooned in his space, isolated by his ideology.

Él and *El ángel exterminador* are exemplars of the convergence of surrealist, expressionist and gothic conventions. The inherent social censure of these traditions is understood and employed by Buñuel and Figueroa in the films and, indeed, in other productions on which they collaborated, to present perspicacious and provocative social critiques. In *Él* and *El angel exterminador* the combination of Buñuel's direction and Figueroa's cinematography creates spaces in which the viewer can experience the dissection and visual extermination of the bourgeoisie.

Examination of the collaboration between surrealist Buñuel and expressionist Figueroa throws new perspectives on the structure of Mexican society, its internal workings and psychoses. The concept of the *burgus* as the bourgeois powerhouse, alienation and the use of gothic tradition are areas I am now investigating in relation to Buñuel's work. The representation of class from a cinematic viewpoint is also a realm for further consideration, particularly the ways in which class is represented visually and the effects of transnational collaborations and coproductions on the representation of class in diverse cultures.

Figueroa worked with Buñuel on seven films over a fifteen-year period. It was the most consistent collaboration with a director of photography that the director had. Indeed, he publicly cited Figueroa as his preferred cinematographer (Poniatowska: 106). Throughout their collaboration, they both challenged dominant Hollywood conventions to develop a style that subtly complemented the subversive narratives of the films. Figueroa's meticulous manipulation of expressionist technique, developed the conventions of a Mexican cinematic Gothic, to visually expose Buñuel's recurrent themes of isolation, social displacement, exile and madness and set a cinematic paradigm for subsequent filmmakers, "You keep on looking for me in the past. But I'm here in the present" (Edith to Thomas in *Crimson Peak* (del Toro, 2015)).

The Mexican Gothic vision, developed by Figueroa and Buñuel in the past, is indeed very much here in the present, no more so exemplified but in the work of the self-confessed exile, Guillermo del Toro and his work with cinematographers Guillermo Navarro and the Danish Dan Laustsen.[8] Following filmmakers, including Alejandro Jodorowsky, Carlos Enrique Taboada[9], Miguel Morayta,[10] Juan Lopez Moctezuma, who went on to further Mexican cinematic Gothic, del Toro, in collaboration with Navarro and Laustsen takes the visual language into the digital age and out to a new, contemporary global audience.

Del Toro brings the Gothic, so integral to the Mexican experience expressed by Figueroa, into a contemporary foreground.

> It would be a cliché to say because I am Mexican, I see death in a certain way. But I have seen more than my share of corpses, certainly more than the average First World guy. I worked for months next to a morgue that I had to go through to get to work. I've seen people being shot; I've had guns put to my head; I've seen people burnt alive, stabbed, decapitated because Mexico is a very violent place. So I do thinks that some of that element in my films comes from a Mexican sensibility. (Del Toro in Kermode: 2006)

The parallel, interweaving narratives of *Pan's Labyrinth* and the violence of *Devil's Backbone*, both shot by Navarro, point to the Gothic duality of this Mexican experience; the tensions between a brutal, external "reality" and an internal "imaginary" which is as resonant in the characters' lives as the real worlds they inhabit. Del Toro names his top three favorite films as *Pan's Labyrinth* (2006), *The Devil's Backbone* (2001) and *Crimson Peak* (2015), the most personal and Gothic examples of work. (Diestro-DoPido: 2015, 6)

Speaking of his long-standing professional relationship with Navarro, with whom he shot *Pan's Labyrinth* and *The Devil's Backbone*,[11] del Toro says,

> he's not afraid of the dark, I've always fought for movies to be as black as possible, with figures emerging from the darkness. (Calhoun: 2007, 3)

His remarks are reminiscent of Karl Freund's visual mantra, cited above, where the creation of shadows is more important than creating light.

Controlling the light to emphasize the dark areas of the frame creates a depth to the image and gives the audience an uneasy awareness of "other" unseen, elements; darker, invisible parts of the visual narrative that the viewer experiences in their imagination. Buñuel and Figueroa worked with this tension in their work together and the working relationship of del Toro and Navarro echoes this with their confidence in letting the images carry the narrative and themes. Of *Pan's Labyrinth*, Navarro says, "This isn't really a dialogue-based movie – the images become the grammar of the story" (Calhoun, 2007: 1). Whereas Buñuel and Figueroa worked in black and white, del Toro has developed a Gothic color palette with both Navarro and Laustsen. Color is fundamental to the visual language and del Toro created a color code for the camera, production design and costume departments (Calhoun, 2007,1 and Diestro-DoPido: 2015, 6).

Set in Fascist-ruled Spain during the 1940s, *Pan's Labyrinth* enters the world of young girl, Ofelia. She and her mother are taken by her new step-father, a captain in Franco's army to a an outpost where he is to eradicate the remaining socialist rebels. Ofelia enters a fantasy world which is as potent as the unimaginable cruelty of the film's real world. Through a mythic quest she confronts her own internal and external monsters, supernatural and human.

The fantasy world of Ofelia and the Faun is primarily crimson and amber, the warmth of which is carried over for the scenes with the rebel fighters. The world of Vidal and the fascists is blue and green. However, del Toro and Navarro wanted to bring the dark, political, emotional and psychological complexity of the story to the fore, to show the way in which the internal and external worlds of the story overlap and inform each other. This they accomplished by mixing the color key codes until as Navarro states, "we found the language we needed to help the audience understand the complexity of the movie" (Calhoun: 2007, 1).

The film was shot of 35 mm and Navarro used three Kodak stocks for the film. Vision 250D 5246, Vision2 500T 5218 on Moviecam Compact cameras and an Arri 435ES for the visual effects scenes. The Vision2 200T 5217 was used for the blue palette and day for night sequences in the forest. This was achieved by underexposing by 3-4 stops to capture the point

> [w]here the worlds are crossing each other. You feel in a naturalistic environment of the night but with certain ingredients that sunlight gives you. We were

> basically collecting light with boards and silver reflectors, so that when the sun hit the greens in the forest, they would pop. (Calhoun: 2007, 2)

The color coding in *Pan's Labyrinth* is evident and consistent, the blue-silver coldness and amber warmth expressing the psychological, political and emotional themes in the narrative. With the latitude of the stocks and lenses he used, Navarro was able to minimize lighting to concentrate on the textures and darkness in frame creating a gothic resonance throughout the film. The use of small lamps and minimal sources and relying on "bouncing" light in scenes enabled greater flexibility for the sense of roaming camera that has become del Toro's trademark.

The movement draws the audience into the scene, engaging them in a close visual relationship with the film's world and the characters.

Navarro's personal reflection on the film (which earned him an Oscar) is significantly Gothic, expressing a desire to transgress, "disobey" and be "wrong," saying that

> We took enormous chances on this movie. It's a movie about choices and disobedience. Disobedience puts you at risk, but it's also what you need; you cannot just follow the rules, you need to take chances. We adopted that as our rule. I did things on this film that I've never done before and that in another context would have been interpreted as wrong. (Calhoun 2007: 4)

In his latest feature *Crimson Peak* (2015), del Toro develops the Gothic legacy of *Pan's Labyrinth* (2006) further, with cinematographer, Dan Laustsen, innovating on gothic tropes with the use of visual and digital effects and a meticulous consideration of framing, lighting and, of course, color. The mise-en-scene was central to del Toro's vision,

> I wanted to build the house, I wanted to do the wardrobe, because I wanted to hide several clues of the story and the characters in the wardrobe and the sets. I wanted to make it like a living painting. (Mar Diestro-DoPido: 2015, 2–3)

Set at the turn of twentieth century in North West England, Crimson Peak is the site of Allerdale Hall, a crumbling mansion. Young American author, Edith Cushing is transported to this family seat by Sir Thomas Sharpe who with his sister, Lady Lucille are the heirs to the Sharpe bankrupt family fortune and industries. Edith soon realizes that her husband is not what he seemed and once ensconced in the mansion, ghostly happenings and supernatural entities reveal the true story of the family and its secrets.

The film's narrative revels in the dualities of the Gothic, Edith the heroine is trapped on the cusp of the nineteenth and twentieth centuries, migrating from American bright, light modernity to Old European (British) dark and shadowy tradition, the rise of the new bourgeoisie and the demise of ancient aristocracy. The eponymous Crimson Peak and its mansion, Allerdale Hall, is a transnational, trans-temporal, trans-social and acutely transgressive space. The house is the *burgus*, the seat of power and obsession, as are the mansions in *El angel exterminador* and *El*. But del Toro pushes the trope further, the house is a character in itself, as director states, "The house is a creature, it is a monster" (Pond, 2015,1)

> I gave the house windows that look like eyes and the house almost becomes all those ancestors watching over them. (del Toro, interview with Mar Diestro-DoPido: 2015)

Del Toro looked toward the classics of US Gothic cinema when he was developing the visual language for the film, *Rebecca* (1940, Hitchcock DP Barnes) *Jane Eyre* (1943, Stevenson, DP Barnes) and *Great Expectations* (1946, Lean, DP Green). Significantly, all of these films were shot in black and white.[12] He cites *Wuthering Heights* (Wyler, 1939), whose DP was Gregg Toland, as his first influence and as discussed above, Toland, was Figueroa's highly influential mentor. I would suggest that where these seminal cinematographers and directors used the chiaroscuro of light and dark and pushed the film negatives in the lab to achieve high contrasts through a range of grays, to communicate duality, conflict and opposition, del Toro and Laustsen use a saturated color palette that encapsulates and exteriorizes the internal workings of the characters, to give life to the house itself and provide a color code for traditional gothic tropes.

> I wanted the past to be represented by the colour red, which is the clay of the house and at the same time is the blood of the earth. It is about these aristocrats effectively sucking the blood out of the earth until it dries and even then refusing to let it go. The movie is divided into two, a golden passage which modernity and then a cyan blue passage which is Allerdale Hall. In the middle you have the colour red linking Lucille to the ghosts, to the ground. (del Toro, interview with Mar Diestro-DoPido: 2015)
>
> Guillermo's approach was to do something very colorful and very dramatic and go old-fashioned and I'm very much into this deep, deep shadowing but you still see the set, the wardrobe and the actors. And we went back to the cold steel blue of *Mimic* for the nighttime and matching that with the very warm amber of the faces. (Lauststen, interview with Bill Desowitz, 2015)

Rather than using CGI, del Toro and Lauststen worked with handcrafted props, special effects make up and specially constructed sets. (Diestro-Do Pido: 2015, 4) Perspectives change during sequences through changing prime lenses and shooting at wide apertures to capture flickering candles and deep shadow. Digitally adjusting color, Lauststen capitalized on the contrast of blues and warm candlelight within the same frame to create the tension of

Edith's personal situation, isolated externally from her physical space while struggling with her internal alienation;

> We're using the shadows and you're feeling the reflections of the house. She's not alone but alone in her soul. And, of course, then she's running through the house and going out the front door and there's a big storm out there. There's no moonlight but, again, steel blue atmospheric light. (Lauststen, interview with Bill Desowitz, 2015)

The use of kinoflo lights rigged up on the chandeliers and on the ceiling provided control of the lights to allow 360 degree camera movement in some scenes, notably in the ballroom sequences. The blue is maintained by backlighting through the windows. The freedom of movement enabled del Toro and Lauststen to experiment with angles, many of which are out of kilter on dutch tilts. The visual language for the film exactly mirrors that of Figueroa and Buñuel in their work together; shedding light only on what the audience needs to see, to maintain a tension between the visible and the unknown present in the darker areas of the frame to access the audience's visual imaginary and suggest dark, invisible and "other" presences.

The legacy of Buñuel and Figueroa, in true Gothic fashion has lived on, transformed and transmuted by subsequent filmmakers. Del Toro, the most prominent heir apparent of the Mexican Gothic tradition, together with his collaborators takes up where the two most fervent critics of the bourgeoisie, Figueroa and Buñuel left off, ironically carrying on a visual lineage and legacy of which any bourgeois would be proud.

NOTES

1. Poniatowska, 1996: 106

2. For example, Horace Walpole, the "father" of the Gothic, situated his first novel *The Castle of Otranto* (1764) in an ambiguously located, imaginary place. He credited the first edition as a translation of a 1529 manuscript he found in Naples, written by Onuphrio Muralto, Canon of the Church of St. Nicholas at Otranto and used the pen name of William Marshal. Mary Shelley, created her monstrous doctor, *Frankenstein* (1823) in Switzerland and the first feature length film version was brought to cinema screens in Hollywood by British director, James Whale, with English actors Colin Clive as Dr. Frankenstein and Boris Karloff as the monster. The sequel, *Bride of Frankenstein* starred Elsa Lancaster, again British. Whale who was openly gay identified himself as "other" and was heavily influenced by German expressionism. For example, he screened Wiene's *Caligari* constantly during the production of *Frankenstein*.

3. See, for example, Johnson, M. Sihvonen, T. (2008) "On the Dark Side: Gothic Play and Performance in a Virtual World" in Virtual Worlds Research Vol.1 No.3.

4. See also Glennis Byron's Introduction in *Globalgothic*. Manchester, Manchester University Press: 2013, pp. 1–10

5. In his acceptance speech for the National Arts Prize in 1971, Figueroa said "In transfiguring reality with a mechanical instrument, reality transformed me and made me grow as a man among men" (author's translation). (Al tranfigurar la realidad con un implemento mecanico, la realidad me transfiguraba a mi mismo y me hacia crecer como un hombre entre otros hombres.)

6. See Buñuel (1984: 201–202) for a discussion of the initial criticism in Mexico and the subsequent approval. See also Evans (1995: 73)

7. Buñuel cited in de la Colina and Perez Turrent (1992: 125)

8. Del Toro's decision to exile himself from Mexico after the kidnapping of his father in 1997 is well documented. See for example Kermode (2006).

9. For an insightful analysis of the Gothic and Taboada, see Kerry Hegarty's "Female Specters: The Gothic Horror Films of Carlos Enrique Taboada" in *Flow. Journal*, Vol 19 no, 7. http://www.flowjournal.org/2014/02/female-specters-carlos-enrique-taboada/#footnote_0_22176. Accessed July 16, 2016.

10. Miguel Morayta worked with Figueroa on *La Casa Colorada* (1947)

11. Del Toro and Navarro have also collaborated on, *Cronos* (1993), *Hellboy* (2002), *Hellboy II: The Golden Army* (2008) and *Pacific Rim* (2013)

12. Del Toro, interview with Mar Diestro-DoPido: 2015.

BIBLIOGRAPHY

Acevedo Munoz, E. (2003) *Buñuel and Mexico. The Crisis of National Cinema*. Berkeley: University of California Press.

Botting, F. (1996) *Gothic*. London: Routledge

Buñuel, L. (1984) *My Last Breath*. London: Vintage.

Castle, T (1995) *The Female Thermometer, 18th Century Culture and the Invention of the Uncanny*. New York: Oxford University Press

Calhoun, J. (2007) 'Fear and Fantasy, Pan's Labyrinth' *American Cinematographer*, January 2007. https://www.theasc.com/ac_magazine/January2007/PansLabyrinth/page1.php. Last accessed July 17, 2016.

Degiglio-Bellemare, M. Ellbe, C. Woofter, K. (eds) *Recovering 1940s Horror Cinema*. New York: Lexington Books

De la Colina, J. & Perez Turrent, T. (1992) Conversations with Luis Buñuel. (trans. Paul Lenti) New York: Marsilio Publishers.

Diestro-DoPido, M. (2015) "Ghost Hunter" *Sight and Sound* Vol. 25 (11) pp. 22–26

Evans, P. (1995) *The Films of Luis Buñuel. Subjectivity and Desire*. Oxford: Clarendon Press.

Forshaw, B. (2013) *British Gothic Cinema*. London: Palgrave MacMillan

Fuentes, V. (2004) "Confluences: Buñuel's Cinematic Narrative and the Latin American New Novel in *Discourse*, 26.1 and 26.2 Winter and Spring. Detroit: Wayne State University Press pp. 91–110

Higgins, C. (2004) "Exterminating Visions: The Collaboration of Gabriel Figueroa and Luis Buñuel" in Santaloalla (et al eds) *Buñuel siglo XXI*. Zaragoza: Prensas Universitarias de Zaragoza pp. 213–220

Higgins, C. (2008) *Gabriel Figueroa: Nuevas Perspectivas*. Mexico: CONACULTA.

Kaye, H. (2012) "Gothic Film" *A New Companion to the Gothic*. Ed. David Punter (Oxford: Blackwell) 239–251

Kermode, M. (2006) 'Pain should not be sought—but it should be avoided' in The Guardian 5.11.2006 https://www.theguardian.com/film/2006/nov/05/features.review1. Last accessed July 17, 2016.

Kristeva, J. (1991) *Strangers to Ourselves* (trad. LS Roudiez) New York: Columbia University Press.

Langenbach, J. (2010) The Gothic and the Phantasmal screen: kindred spirits and their journey from East to West. In *Gothic N.E.W.S* Ed. Gilles Mengaldo. Paris: Michel Houdiard

Lundemo, T. (2009) "In the Kingdom of Shadows: Cinematic Movement and Its Digital Ghost" in *The YouTube Reader* eds. Pelle Snickars and Patrick Vonderau. (Stockholm: National Library of Sweden) p. 316

Och, D. Strayer, K. (2014) *Transnational Horror Across Visual Media*. London: Routledge

Paz, O. (1961) "El poeta Buñuel." *Nuevo Cine* No 4–5. Mexico City

Peters, J.D. (1999) *Speaking into Air: A History of the Idea of Communication* (Chicago. University of Chicago Press)

Piatti-Farnell, L. Beville, M. (2014) *The Gothic and the Everyday: Living Gothic*. London: Palgrave Macmillan

Pond, S. (2015) "*Crimson Peak* Director Guillermo del Toro on How Hollywood is Limiting Horror Movies." The Wrap 13.10.2015 http://www.thewrap.com/crimson-peak-director-guillermo-del-toro-on-how-hollywood-is-limiting-horror-movies/. Last accessed July 17, 2016.

Poniatowska, E. (1996) *La mirada que limpia*. Mexico: Diana

Punter, D. (2012) *A New Companion to the Gothic*. Ed. David Punter. Oxford: Blackwell

Miles, R. (2008) Gothic Technologies: Visuality in the Romantic Era https://www.rc.umd.edu/praxis/gothic/intro/miles.html

Paz, O. (1971) "El poeta Buñuel" *Las peras del olmo* Barcelona: Seix Barral.

Risner, J. (2014) 'Filmic Spaces in Contemporary Latin American Gothic Horror Cinema' in *Studies in Gothic Fiction*, Vol. 3 issue 2 http://studiesingothicfiction.weebly.com

Sears, R. & Pocknell, D. (eds) (2003) *Making Pictures. A Century of European Cinematography*. London: Aurum Press

Said, E. (2000) *Reflections on Exile and other Literary and Cultural Essays*. London: Granta

Williams, A. (1995) *Art of Darkness: A Poetics of Gothic*

Wood, M. (1993) Buñuel in Mexico. In King, J. (et al., eds) *Mediating Two Worlds, Cinematic Encounters in the Americas*. London: British Film Institute

Chapter 9

Gravity and the "Lighting Designer" Controversy

Cinematographers, Special Visual Effects Artists and the Rhetoric of Digital Convergence[1]

Julie Turnock

The release of *Gravity* (Cuarón, 2013, Director of Photography, [DP] Lubezki) created a forum for a number of debates, popular, professional and academic, about the role of digital technology in filmmaking. Academics wondered, since most of the film, excluding the actors' faces, was digitally generated, should it be considered a live action film or an animated film? For cinema professionals, the production gave rise to other controversies: if the majority of the film was designed and rendered in the computer, where do we divide the responsibilities or assign credit for its excellence? For cinematographers in particular, *Gravity,* in the wake of similar acclaimed, Academy Award-winning films such as *Avatar* (Cameron 2009, Fiore credited DP[2]), *Hugo* (Scorsese 2011, DP Richardson), and *The Life of Pi* (Lee 2012, DP Miranda) emerged as a flashpoint for a debate on what is the job of a cinematographer on nearly fully animated Hollywood blockbusters. As Roberto Schaefer asserted:

> To me *Gravity* was an animated movie. Beautiful! Beautiful visual effects. Absolutely stunning. It should have gotten every award for visual effects ever designed, ever given. But it should not have gotten the Best Cinematography award.[3]

The *AC* feature article on *Gravity* weighed in on the controversy by proposing the idea of the DP as the "lighting designer." As the author put it:

> [*Gravity*] is the fruit of a five-year collaboration involving director Afonso Cuarón, Emmanuel "Chivo" Lubezki, ACS, AMC, visual effects supervisor Tim Webber and their talented teams. [. . .] In many ways *Gravity* provides a new paradigm for the expanding role of the cinematographer on films with significant virtual components.[4]

The lighting designer, as modeled by Lubezki's work on *Gravity*, would go beyond his traditional role as "painter of light"—that is artistic shaper of light on-set—and would direct this specialized expertise to other levels of production down the pipeline.[5] With the increasing reliance of big budget Hollywood releases on digital special visual effects,[6] one might think the lighting designer designation was proposed as a way to stem the encroachment of effects workers onto cinematographers' territory. However, I argue that the opposite is true: the lighting designer role is a way to distance the DP's work from effects workers, so that cinematographers' and effects workers' territories do not get rhetorically confused or conflated.

Moreover, in attempting to bolster the position of the cinematographer in light of fluctuating professional categories, the *AC* article also inadvertently reinforced a number of myths and misunderstandings about digital technology encouraged by what Tom Schatz calls "conglomerate Hollywood": that digital technology unproblematically facilitates fluidity, interoperability, interchangeability and increased ease of collaboration.[7] In Hollywood's contemporary post-Fordist creative labor economy, the various rhetorical positions around the *Gravity* production aim to shape perception about industrial conditions in order to impact actual working conditions. A close analysis of the rhetoric of this *AC* article and others within an historical context has important consequences for unmasking corporate rhetoric about digital technology as an attempt to amass all below-the-line digital workers into the same undifferentiated pool, thereby neutralizing their powerful labor protections. Effects workers, as the vanguard of cinema labor shifting to digital technologies in the early 1990s, did indeed see their labor protections erode as they transitioned from cameras to computer monitors. Moreover, this chapter demonstrates how Hollywood corporate rhetoric around digital interoperability permeates the industrial discourse to the "below the line" technical artists like cinematographers and effects artists, who feel they need to tow the producers' line in order to be seen as cooperative in this increasingly precarious creative labor context.

The first quote that leads the chapter from the *American Cinematographer* 2013 feature article on *Gravity* suggests what an expanded role of the cinematographer to serve as the production's "lighting designer" would mean. The lighting designer would lead the technical look of the film from start to finish. The article continued to describe the lighting designer's responsibilities in this so-called "new paradigm."

> By all accounts, [DP] Lubezki was deeply involved in every stage of crafting the real and computer generated images. [. . .] He created virtual lighting with digital technicians, lit and shot live action that matched the CG footage, fine tuned the final rendered image, supervised the pictures conversion from 2-D to 3-D, and finalized the look of the 2-D, 3-D and IMAX versions.[8]

Presenting a model for an ideal collaboration, the article heavily credits teamwork among Cuarón, Lubezki, and visual effects supervisor Webber (of UK-based effects company Framestore) as the key to the film's success. The smooth rhetoric of teamwork in articles such as this and others suggest that with contemporary digital technology, this type of coordination closely integrating directing, shooting and effects across the production workflow from previsualization to final cut is already routine. However, as the second quote by cinematographer Roberto Schaefer, frequent collaborator of directors Marc Forster and Christopher Guest, suggests not only is *Gravity* a rarity among productions, the questions its model raises are far from settled. Questioning these surface statements is necessary to oppose the bland corporate discourse the industry deploys at all levels to create the illusion of a trouble free, easygoing, even fun production. And in fact, this mild, PR-approved discourse covers over a number of disputes. For example, outside of pre-release promotional material, one can find grumbling in both cinematography and effects communities over Best Achievement in Cinematography Academy Awards given to work consecutively on *Hugo* (2011) *The Life of Pi* (2012) and *Gravity* (2013), as well as how to properly categorize and credit films with extensive digital environments. Christopher Doyle, HKSC, prominent international DP and frequent collaborator of Wong Kar Wai spoke out in frank terms against the animated environments in *The Life of Pi* saying, "it's a f—ing insult to cinematography," and more "It's not cinematography."[9] Effects Artists were equally displeased, but because they did not feel properly credited for the extent of their contribution on the film. As effects worker and employee of Rhythm and Hues Bruce Branit put it, "Neither Ang [Lee] nor his winning cinematographer, Claudio Miranda felt they needed to thank or even mention the VFX artists who made the sky, the ocean, the ship, the island, the meerkats and oh yeah . . . the tiger."[10] The crack in PR discourse indicates a push back against the corporate homogenizing of digital work, and instead articulates important artistic creative differences among different kinds of specialized digital labor.

Not surprisingly, the public discourse that surrounds contemporary filmmaking must always be seen as a tool of promotion, whether on the cover of *People* magazine or in the pages of more specialized publications like *American Cinematographer*. Therefore, this is not going to be an essay about the production *Gravity* or *Life of Pi* as it "really" happened, but rather about the discourse that surrounds them and other effects-heavy films, and how this discourse creates and perpetuates the very misconceptions about which the cinematographers like Schaefer are complaining. Instead, I will analyze the constructed public rhetoric attached to *Gravity* that brings out the surprising ways that corporate rhetoric from above permeates professional discourse, and through it, obscures the actual working practices professionals engage in. *Gravity* and other recent effects-heavy movies that straddle the line between animation and live action, require us to ask, "who paints the light?"[11] and helps to frame the rhetoric around historical conflicts and the stakes of professional categorization brought on and seemingly threatened by digital technologies. However, I argue that the blurring of professional lines derives more from corporate strategizing rather than actual technologies.

THE MYTH OF DIGITAL CONVERGENCE

What I am calling the technical myth of digital convergence, the incorrect assumption that digital technology turns "everything into zeros and ones" and is therefore easily interoperable, must be dispelled. What is often referred to as "the digital" in cinema is likewise an equally vague term for technologies that encompass image capture, processing, transmission, and eventually theater exhibition and streaming. This myth creates the false impression that images flow fluently down the production pipeline as they pass through previsualization, design, on-set production, editing, effects, color correction, then merge into a final product and are beamed to cineplexes around the world. As a corporate myth, digital convergence is useful in how it first of all simplifies a number of technologies for easier public consumption, much as Donald Crafton as discussed the public role of the term "electricity" as a way to streamline the public understanding of a number of only very loosely related technologies of the early twentieth century.[12] As any artist with any experience along the production pipeline knows, this effortlessness is far from the truth. Furthermore, skills in one area of digital technology are not easily transferrable to another area. For example, someone skilled in digital intermediate would need a great deal of retraining to switch careers to become a digital character animator, even if they already know how to use Autodesk Maya or other imaging software.

Despite the incompatibility of various proprietary technologies and software, as well as the different skills and specializations of various areas of production, this myth of digital interoperability has also impacted academic special effects discourse. The academic rhetoric on digital convergence rarely discusses the technological particulars. Rather than questioning these assumptions, more often than not, they inadvertently reinforce the myths.[13] Stephen Prince in his most recent book has argued that since digital technology is so widespread in contemporary filmmaking, the term "special effects" in its popularly understood sense as "special" footage in regard to the "normal" unmanipulated footage is outdated and even irrelevant.[14] While I do not disagree with this notion, Prince takes this idea even further by echoing the smooth corporate idiom of digital convergence. He argues that digital technologies across many imaging systems, including what is traditionally understood as visual effects but also including editing, cinematography, 3-D, animation, and also distribution and projection, have become so prevalent that, that all visual digital manipulation should be rhetorically converged into the term "visual effects." This heuristic act, while seemingly useful, unintentionally bolsters Hollywood conglomerates' devaluation of specific artistic skills, including cinematography and effects work, by reinforcing the strategic move to re-categorize them all as unspecialized "digital workers."[15]

Technology is just one way Hollywood production remains as inherently unstable and unpredictable industry as ever. Nevertheless, Hollywood corporate promotional rhetoric has long worked to present itself harmonious and seamless, especially in times of technological and economic change. Hollywood's contemporary discourse about "the digital" and digital convergence is no exception. Early Hollywood labor has typically been described by academics such as David Bordwell, Tino Balio and Douglas Gomery as loosely adhering to early twentieth-century notions of Fordist efficiency organization.[16] The fantasy image of the studio structure promoted by Hollywood and generally embraced by academics, promotes an image of variously specialized worker-cogs doing their individually skilled part in a clear managerial hierarchy.[17] It has long been a powerful image of what Balio calls a "Grand Design" and is more popularly (and temptingly) called a "Dream Factory."[18] Nevertheless, as Don Crafton has suggested, this image is a rhetorical fantasy, where Hollywood studios presented its corporate structure as similar to the auto industry or other kinds of manufacturing, in order for the entertainment business to appear more solid, stable, and safe of a corporate investment.[19] As strikes and other labor and legal action throughout Hollywood history suggest, actual labor conditions were often much more fluid and contentious than the self-image insisted.[20]

In the contemporary context of conglomerate Hollywood, rhetoric promotes fictions and misunderstandings of digital technology to advance a kind of Neo-Fordist image of operations that reproduces buzz words of infinite unboundedness and convenience like flow, seamlessness, and streaming. Instead of the bygone harmonious labor specialization of auto factories, the contemporary Hollywood corporate fantasy of the digital age approaches that of a hip media company, where digital technology enables and produces not specialized workers, but those of maximum flexibility, adaptability and interoperability of skills. In other words, workers who mirror the popular image of "digital convergence" of creative labor.[21] This strategy promotes a more of a reverse-Fordism where (a fantasy of) digital technology would allow all levels of cinematic production to be interchangeable, and therefore that all digital labor and laborers would be easily swapped out. Such a de-specialization of the work force would obviously be advantageous to corporate interests.

OBSTACLES TO DIGITAL CONVERGENCE

Why, exactly, is the assumption of technological interoperability wrong? Theorists such as Lev Manovich have discussed the pixel-level manipulability of digital material since the 1990s.[22] However, while broadly true, in fact, the main obstacle to smooth pipeline digital flow is both technological and cultural. The greater malleability of digital elements in the production has in fact confused the traditional historical linear (pre-production, shooting, post-production) and hierarchical format, but not in the way the myth of the granular manipulability of zeros and ones is usually characterized. It is precisely the how the elements get handled across the pipeline that can create conflict and breakdowns.

In technical jargon, technicians call this "raw" versus "baked" material. Raw files are image information taken directly from the camera sensor, and includes information taken at the time of recording, such as the camera settings and processing parameters.[23] For example, all digital photographs start as raw files, but as such are not "pictures." They must be converted or compressed into a smaller, easier to store file, such as a jpeg in still photography. In the feature filmmaking process, this is more complex, but workers down the digital line rarely work with the raw files DPs capture, and instead, receive data with some information, such as color latitude, "baked" in already, meaning there are aspects of the image that cannot be significantly changed. DP Roberto Schaefer insists each set of hands makes changes that other levels cannot then undo:

> [Visual effects teams] work off the DPX file, which has a certain amount of color information baked into it. You can't stretch it as far as you could your "raw" or your negatives, which has all the latitude there, you can use the top end, the bottom end, you can do windows, you can find all ways to get the image that you want. [. . .] They will bake in a look on the DPX files that when you get it back you can't, you don't have the facility to put it back into the rest of the look. And especially if they are composite files, you get it where the foreground just doesn't work with the background, it's a different color space.[24]

Therefore, as Schaefer put it, the files are baked successively on various levels of production with different software, often proprietary, they become less and less malleable. Schaefer suggests that better cooperation earlier on would allow the layers to bake more evenly, as it were. Therefore these different "digitals" obstruct Manovich's and other's fantasy of perfect plasticity of all digital material.

Barriers are also cultural and historical. This digital convergence discourse seeks to obscure the persistence in actual cinema production of long-standing strict skill specialization. For example, cinematographers respect the professional boundaries and skill sets and do not want to do the job of, say, the production designer or the editor. Moreover, although all are using digital tools in order to do their job, these technologies are no more interoperable than the "analog" cameras, drafting boards and editing tables that preceded them. These strictly divided professional identities and the cultural context of specialization, enforced especially by guilds and unions, make perfect digital interoperability virtually impossible.

Nevertheless, corporate discourse works tirelessly to attempt break down the labor categories that empower unionization and organization. One of the categories conglomerates have been most successful in creating this impression of full technological and labor interoperability is in special visual effects. No production area has been as materially affected by these technological misunderstandings as the effects industry: effects workers have lost labor and guild representation in the move to digital technology, and seen their jobs disperse around the world, as will be discussed later on. Other areas of production, such as cinematography but also production design find themselves the victims of misunderstandings in public discourse around digital convergence, despite the realities of their actual working conditions. I argue these errors obscure the artistic specificity and specializations and through that, their professional status.

Again, what is at stake is not about the erosion of the "actual" considerable skills various Hollywood artists possess. Instead, I argue that Hollywood corporate discourse about what is typically referred to publically as "the digital" purposefully and strategically creates confusion to those outside of the industry. This can impact, for example, lawmakers who pass economic incentives to attract production, educational institutions that create training programs for future workers, or young people attracted to work for low pay just for the "fun" of working on a movie franchise they love. In rhetorically eliding these jobs, and actively creating public misunderstandings about which artist does what and how, the corporate rhetoric of digital convergence creates a slippery slope where cinematographers (and all other below-the-line film artists) can be categorized as "digital worker." This recategorization can be pathway to losing their secure professional identity and status as artists, as has been the case with effects workers.

RHETORIC BELOW THE LINE

The heated language by Christopher Doyle and others reflects many film workers' desire to push back against not only public misperceptions, but also implicitly, the corporate rhetoric that tends be at its root. John Caldwell's ethnographic account of movie professionals' public rhetoric is helpful as a way to characterize how so-called "below the line" professionals both discuss cooperation and characterize the separation of skills among professionals. As he argues, "Understanding 'how tools mean' invariably animate and upset three interrelated registers: labor relations, corporate alignments, and aesthetic practices."[25] As Caldwell acknowledges, this interplay among public rhetorics never creates a picture of "reality," but is always constructed.[26] Moreover, because it is in the below-the-line worker's interest to seem easy to work with, this carefully formulated public rhetoric stifles many of the below-the-line workers' dissent.

These workers' discourse also speaks to the confusion the corporate discourse has created within these long-standing labor categories as professional organizations such as the ASC attempt to redefine themselves along the corporate line. For example, during Richard Crudo's tenure as ASC president, he frequently used language reminiscent of corporate "progress" speak, for example exhorting members to adapt to contemporary conditions or die, telling "nostalgics" reluctant to embrace the changing conditions, "Get over it." And that doing so "strengthens the cinemaographers' position."[27] In this way, I argue that these disputes are at the crossroads of a number of important issues facing both Hollywood's changing corporate structures in light of digital technologies, as well as artists' self-image as they adapt their skills and professional images. And these issues can be analyzed though the interaction of public and private statements as they clash. What these disputes indicate is that the nearly fully digitally animated environments in movies like *Gravity* not only appear to blur the line between long-standing production credit (and award) categories, but that they animate the worst nightmares of many below-the-line workers, especially cinematographers.

Cinematographers would seem to have a lot to lose. Among "below the line" workers, they are typically among the best paid, have a powerful guild to represent their interests, and a long-standing high status as artists on the set. Since cinema's earliest days, they have long enjoyed the specified and irreplaceable artistic position on the set as virtuoso "painters of light," respected for their artistry derived from mastering the physical cameras and lenses on the set.[28] Although some use heated rhetoric, when it comes to cinematographers's public statements on these matters, most take a more conciliatory tone. Many assert this digital dispute is simply a matter of being specific about giving credit where credit is due.[29] If the "cinematography" was created in the computer, and not by the light hitting the lens on the set as captured by the cinematographer and their team, then it should not be considered cinematography. They consistently claim their respect for the post-production work of effects artists and others down the pipeline, but also assert their artistry comes from a different skill set.[30] Nevertheless, while all artists have had their work impacted by digital technologies, the historical relation between cinematographic "camera workers" and effects artists highlight exactly how digital technology, and a skillful manipulation of public discourse encourages this downward labor trajectory. Recounting how effects workers lost their status as artists, as I will do in a later section, as well as how the contemporary context restages this demonstrates that cinematographers ought to fear this seemingly innocuous discourse of cooperation, harmony and progress.

GRAVITY AND THE LIGHTING DESIGNER CONTROVERSY

Returning to *Gravity* and the issue of lighting designer, we can see how the ASC uses rhetoric to model the "proper" working relationship among director, DP and effects worker but also to implicitly tow the corporate line. Again, the term lighting designer appeared most prominently in the *American Cinematographer* 2013 feature article on *Gravity* as a proposed solution for the expanded role of the cinematographer. While as the journal produced by the American Society of Cinematographers generally provides on one hand a straightforward account of the production (and an extension of promotional machine), it nevertheless can be analyzed as a site of discourse, with the intent to produce rhetorical categories, as the lighting designer designation suggests. Again, the lighting designer is proposed as the point person to lead the technical look of the film from start to finish, meaning supervising both the real and computer generated images, planning virtual camera moves, designing virtual lighting with digital artists as well as lighting live action shooting, tweaking the final rendered image, and supervising the film's look in post-production to the final exhibition format.[31] Although in fact a rare and particular production context, the article uses the *Gravity* shoot to present a model for an ideal "digital" collaboration. [32] The article heavily credits teamwork among director Cuarón, Director of Photography Lubezki, ASC, AMC, and visual effects supervisor Tim Webber (of UK-based effects company Framestore) as the key to the film's excellence. The smooth rhetoric of teamwork in articles such as this and others suggest that with contemporary digital technology, this type of coordination closely integrating directing, shooting and effects across the production workflow from previsualization to final cut is typical of such productions. In fact, articles produced by professional organizations like the ASC rhetorically dodge the sensitive fact that most DPs do not supervise the look of the film as it is finalized in post-production, because he or she has moved on to another job and is typically not paid for post-production work.[33] Further, as has become more and more common, especially since the 2000s as blockbuster films require more and more effects shots, the effects team must rush to finish their work in extremely shortened deadlines, with little input from either the director or the DP. In the industry, this has been called the "War of the Worlds" effect, referring to the shortened effects timeline that was successfully completed on 2005's *War of the Worlds* (Spielberg), and greatly accelerated the post-production timeline for many effects houses.[34] The shortened effects completion timeline on this film was required by specific circumstances, but once the industry saw it could be done, as VFX supervisor on the film John Knoll put it, it made everyone insist the shortened timeline.[35] A lighting designer would streamline this unfavorable situation and bring the effects team in earlier in the production pipeline. Advancing the idea of the expanded role of the DP as "lighting designer" is a way to produce a category that does not yet exist, but benefits the role and status of the DP.[36]

The *AC* article addresses the category misinformation around digital technologies aggravating DPs by recasting it as a positive. By casting the DP the key to the overall look of the production, the article presents the *Gravity* collaboration as an attractive model of frictionless efficiency flow down the pipeline resulting in acclaim, awards, and profits. The article not only presents Lubezki's work on *Gravity*, again, in actuality a production rarity, as an ASC-approved model for the expanded role of the cinematographer as the point person for all technical areas of the film's look. It also suggests how this is an opportunity for DPs to recast their role as even more central to the production.

And Lubezki is active in this redefinition. In case the film's extensive animation created any doubts in the *AC* reader about the DP's central role, Lubezki puts it pointedly:

> I was doing my work as a cinematographer on *Gravity*. In the process, I had to learn to use some new tools that are part of what cinematography is becoming. I found it very exciting.[37]

In other words, a shift to lighting designer for cinematographers should not be resisted or feared, but embraced as strengthening the DP's position. However, the term "lighting designer" and the motivation behind it has not been positively received by all cinematographers, but for perhaps unexpected reasons.

Much academic discourse on the cinematic shift to digital technology has centered around the issue of the "index": the real or imagined direct physical impression—"it was there"—of the world created when light passes through a lens onto a film emulsion.[38] For many theorists of the mid-twentieth century, most notably André Bazin, this ontological connection to "the real" (which some believed in more literally than others) and the filmmakers' artistic shaping of the real, is what made the cinema a new modern art on a par with or superior to the traditional plastic arts such as painting or sculpture.[39] While much academic discussion has focused around parsing what happens to the index in the transition to the digital age, this ontological debate along academic terms has not troubled filmmakers, especially cinematographers. Like the rest of the industry, cinematographers spent the 1990s and early 2000s debating the aesthetic qualities of digital versus film technologies, until c. 2010, when a critical consensus built (though there are still of course dissenters, such as Christopher Nolan's frequent collaborator Wally Pfister, ASC, BSC) that digital cameras had reached an acceptable technological aesthetic parity with film cameras.[40] And again, although there are outliers, virtually every cinematographer working on Hollywood films in the 2010s shoots digitally. Digital capture and film capture may have different aesthetic parameters and potentials, but the cinematographers' molding and "painting" with physical set or location light onto a light sensitive medium (whether film emulsion on chip) remains the center of their artistic and professional identity.[41] In other words, rather than understand digital technology as a severing of the connection between "the real" and the virtual, cinematographers generally treat digital cameras as functional equivalents of film cameras.

This long-ingrained professional identity of cinematographers as painters of light is central to the reason that despite the enthusiasm of Lubezki and by extension, a strong faction of the ASC, cinematographers are very mixed at best on the prospect of DPs as "lighting designers." Moving lighting design into the digital realm potentially thrusts them even further away from in-camera lighting on principal photography and down the digital production pipeline. As already mentioned, Christopher Doyle, HKSC, considers this expanded definition of DP as "not cinematography." Doyle asserts an artistic distinction based in DPs unique skill in translating "real" light artistically in the camera. A lighting designer, who may or may not ever touch a camera, appears to throw the source of this artistry into doubt. With effects artists "painting with light" in the computer and art directors and editors also manipulating the image along the way, what does their credit as director of photography mean? And Doyle is not the only cinematographer who sees the problem in who has control of the image and how it is formed. In an interview with the author, Roberto Schaefer strongly agrees with Doyle on what constitutes the work of a cinematographer: "I agree that [Miranda's work on *The Life of Pi*] was an admirable job, it wasn't an easy one, but to me I agree it was not worthy of even being nominated for best cinematography."[42] Resisting the label "lighting designer" Schaefer draws a line:

> If somebody asks me to come in and give consultation on an animated film, I would be happy to do it. But I wouldn't consider it at all in any stretch of the imagination my cinematography. To me, it's just, I consulted on a look. [. . .] It's a different job at that point. It's visual consultant or visual designer or whatever.[43]

At the heart of the controversy for Doyle and Schaefer and many cinematographers is that the recognition of their specific camera and lens based training, as well as their experienced-based virtuosity to capture real light, is an important part of justifying their high position in the creative hierarchy.[44]

However, it is important to state that this is a hotly debated issue, and not everyone agrees with Doyle and Schaefer. Cinematographer Dean Cundey, ASC, DP on effects-heavy films such as *Jurassic Park* (Spielberg, 1993), *Who Framed Roger Rabbit?* (Zemeckis, 1988) *Back to the Future* (Zemeckis, 1985), and *Apollo 13* (Howard, 1995) takes a more sanguine view. Cundey, tending toward the Lubezki/*Gravity* model in supporting a more expanded role for the cinematographer puts it,

> There is a place for the cinematographer to still be in charge of the visuals in the digital world. And that's one of the reasons it behooves us as cinematographers to understand that new world, not reject it, not be afraid of it, but embrace it.[45]

Cundey is happy to imagine the DP as a kind of "visual technology artist" who translates the ideas of the director into digital form, which then seeps through the production:

> One of the things I talk about often to films students is that cinematographers are among other things the bridge between the science and technology of film and the artistry—the vision of the director. And how do you apply the science to the artistry? So it's not just about lighting a set, there is an awful lot of other things about lenses and camera positions and things like that that are all part of what the cinematographer does to capture and make decisions technical and artistic, make decisions that are what the film should be or is. So I think just to say that the cinematographer would be the lighting designer is possibly true on a couple films.[46]

In other words, Cundey sees as secure the DP's role as visual interpreter of the story, and that the experience of cinematographers have of "real light" is vital to the new context. As to the idea of a professional role of a "lighting designer," Cundey is open:

> Even if half of the movie I was working on was conventional photography and the other half was digital. I would welcome the concept that I would be also a major force in the digital world. [. . .] We should all be engaged in it.[47]

Believing the key is cooperation among departments, Cundey believes that means, "everybody's just working to the same ends as opposed to working on their own."[48] What Cundey calls "going into business for oneself," or working at odds with other phases of production.

As a guild membership point of contention, and a clear indication of the real issue at hand, both Cundey and Schaefer brought up unprompted in the interview the election of Pixar animator Sharon Calahan as a full ASC member.[49] Cundey saw Calahan's membership as a positive step in furthering collaboration along the production pipeline by bringing in artists not traditionally considered cinematographers:

> [Sharon Calahan's] function [on *The Good Dinosaur*, Sohn, 2015] was lighting the shots, lighting the scenes, lighting the whole movie. And you looked at it and said "well, that's the same stuff we do except it doesn't exist in real life." But her sensibilities, how she created moods, how she created pools of light that the characters went through, was all about creating the same kind of mood and emotional reaction in an audience as we do and did. Using physical equipment. And we brought her in. We said "oh, yes this is the way of the future." I think we need to pay a little more attention to that because if we can say here are people whose job it is in an artistic creative way to create images that are . . . serve the purpose of the storytelling, then those people should be brought in so that they can influence how we think of it and also we can influence how they interpret a story or whatever.[50]

When asked if her membership proved controversial, Cundey replied that although there were some qualms about there being no camera involved, that the art of lighting and an appeal to progress won out: "Suppose we never had cameras, we would be doing the same thing." And everyone said 'oh, I guess it's time.' So it wasn't a hard argument as I remember it."[51]

Schaefer had a different reaction to Calahan's membership, "To me, [Sharon Calahan is] not a cinematographer. She is a very talented lighting designer for animation. That's not cinematography."[52] Schaefer instead prioritizes "movies [. . .] that were really crafted by people working with cameras and lights,"[53] not lit in a computer. To Schaefer, the ASC's promotion of a "lighting designer" is acting out of fear of irrelevancy rather than a forward thinking embracing of the future. And one could speculatively see the ASCs position as responding unreflectively to the conglomerate rhetoric of digital convergence that favors the "flexibility" and "nimbleness" of blurred categories.

While one might imagine that this argument is conflict between forces of conservatism versus forces of progress, the lighting designer controversy rather brings forward the material labor consequences to losing a professional core artistic identity through a shift in technology. And while these concerns may seem overblown, historically this professional downward trajectory is exactly what happened to effects workers in the transition to digital technology.

GRAVITY: COLLABORATION WITH THE EFFECTS TEAMS

It is useful to see how the rhetoric of cooperation and convergence is reflected in the *AC* characterization of the *Gravity* production. If the *Gravity* production's enlarged role of the DP as lighting designer is supposed to model an ideal cooperation among director, DP and effects artists, how does that translate to the working conditions of the effects team on the production?

As described by the article, Lubezki laid out an ideal working relationship by suggesting two important shifts in the role of the cinematographer and collaboration among levels of production. The first, vitally, was to redefine the responsibilities and duties of the cinematographer as a "lighting designer" in a mostly digital production. The other innovation of the article was to describe the ideal situation not only for the director and DP's early collaboration with the special visual effects team, but in effect to elevate the visual effects supervisor to near the role of the director and DP.

Many articles had appeared over the years in professional rhetoric in publications like *AC* and *Cinefex* stressing the importance of involving the effects team early on in planning.[54] The early and integrated involvement of the effects team is a long-standing call among effects workers, and nearly everyone would endorse this stance as an ideal. It may seem that the centrality of effects work in blockbuster production would necessitate the close involvement of the effects teams throughout production, but the reality is that (granting exceptions) the visual effects supervisor and the effects teams rarely are involved. Historically, the low status of effects workers at the end of the workflow keep this kind of tight collaboration from happening with any consistency. Effects artists, at their usual back end of the pipeline, become button pushers who (it is often believed) execute the ideas designed by those higher in the hierarchy. They are not regarded as artists who create from their own specific skills. How this low status plays out in the production is that in most large budget films, the producers and/or director plan for effects sequences in previsualization (previs), typically contracted to effects houses such as The Third Floor specializing in previsualization animated storyboards, also called animatics. These plans serve as the blueprints for the effects design. However, the previs company's work or personnel rarely extend past the pre-production stage. Typically, the production hires a primary visual effects supervisor, who may be the lead effects supervisor at one of the contracted effects teams (as is the case with Roger Guyett of ILM on *The Force Awakens*). Alternatively, the visual effects supervisor may not work for an effects house at all, but instead be hired directly by the production company (as is typical with Marvel productions, for example Janek Sirrs on *Marvel's The Avengers*). The visual effects supervisor's job is typically to mete out the effects work, which could be with one or two "lead" effects houses and a handful of pick-up work by smaller houses. Alternatively, as is becoming more and more common, the post-production effects job is often split up among several to many (in the case of *Marvel's The Avengers*, 2012, upwards of twenty) houses. Although some of the effects work is done concurrently with principal photography, most is done well after. Because the effects for a large effects-heavy blockbuster are typically one of the last if not the last job of the production and not completed until the last possible minute, the effects teams are under a great deal of pressure to simply finish the effects on time for the mandated release date. And it is worth emphasizing it is not the effects teams' carelessness or sloppiness that causes the rush to the end, but the more often due to last-minute changes by the production or over-optimistic delivery completion dates. Productions differ, but usually the visual effects supervisor is not involved with the previsualization, and may or may not have been on set or closely consulted during principal photography. And in the post-production rush to the end, there is not a lot of space for collaboration with the director and especially the DP, who after shooting is likely to have moved on to another project.[55]

The *AC* article on *Gravity* promotes an unusually expanded role for the visual effects supervisor as well, albeit under the supervision of Lubezki. First of all, Tim Webber and his team at Framestore were indeed involved with previs, and began generating environments early in the pre-production. The *AC* article narrates the effects team's involvement: "The filmmakers [meaning Cuarón and Lubezki] began their prep by charting a precise global trajectory for the characters over the story's timeframe, so that Webber and his team could start creating the corresponding Earth imagery."[56] And "next the filmmakers defined the camera and characters positions throughout the story so that animators at Framestore could create a simple previs animation of the entire movie."[57]

The intricacies of the project's nearly fully CGI generated settings meant this previs also involved the effects team in what Lubezki called the "prelight." The highly unusual procedure of the DP designing the prelight with the effects team set the look for the overall production, from pre- to principal to post-production. Paul Beilby, a CG lighting supervisor at Framestore reinforces the novelty of this approach, "We worked directly with Chivo." And this process was much more entailed than effects lighters typically get with the production DP, with about a dozen "digital gaffers" working together on the prelight.[58] The VFX producer Charles Howell describes this arrangement as to the production's advantage:

> Because these shots had to be mapped out from day one, many of the lengthy shots didn't really change in the three years of shot production. Because we did a virtual prelight with Chivo, the whole film was essentially locked before we even started shooting.[59]

In this way, the article puts forth an ideal of presenting a mock-up of the entire film in pre-production, as the Framestore effects team had a director and DP-approved blueprint for the post-production rendering and refining of the effects. Therefore the effects team could work parallel to the production, theoretically creating a model that would eventually eliminate post-production.

The article includes Webber in the devising of realms typically the purview of the director and DP. In another unusual move, Webber and his Framestore team also collaborated with director and cinematographer on both designing and conceptualizing the camera moves. The effects team was also vital in what Lubezki's "revelation": what he calls his "LED box" helping to creating naturalistic lighting on the actors faces on the set, to be composed later in the computer. He described it as building a set of boxes out of LED panels and then lighting actor's faces with the previs animation.[60] He had Webber and his team realize this idea by building a twenty foot cube of LED screens that would project the virtual environments across the ceiling and the actors' faces, which also helped the principal photography keep track of the complexities of the lighting sources as the camera moved and time passed. Again, what is usual about this is that the LED box executed by the effects team was built exclusively to light the actors' faces, using the effects-generated lighting. This means the effects lighting led the on-set lighting, not the other way around. This teamwork was considered vital to avoiding the disconnection between the actor's lighting and the animation lighting that happens when the filmmakers do not communicate effectively with the effects team.

Finally, as a finishing element in post-production, the article describes Lubezki's involvement in finessing the final rendered effects images. Lubezki, for example, suggested to Webber to "add lens flares and chromatic aberrations in the CG so it would look as if the [entire] image had been captured with a camera."[61] Again, we see a rare example of the DP collaborating with the effects team at the end of post-production, and another way for Lubezki to assert his visual authority. Though it seems to be an example of the smooth language of cooperation, Webber nevertheless makes the hierarchy clear:

> It was great working with Chivo, who is not only an incredible talent but also very willing to use this new technology and embrace lighting within this new environment. Even though very little of the film is physically lit in the way Chivo would normally do it, his touch is all over everything.[62]

Even the article's author Benjamin B throughout constructs the article as a model tight collaboration among team players, with paragraph structure reinforcing the point by weaving Webber into the account early and often. Nevertheless, the rhetoric also reinforces a divide between "the filmmakers" (Cuarón and Lubezki) and "the effects team" or "animators," emphasizing Webber's importance, but subordination to the directives of the director and DP.

With the DP the key to the overall production, the article presents the *Gravity* production as an attractive model of tension-free efficiency resulting in acclaim, awards, and profits. However, not surprisingly left out of the account are a number of important controversies manifesting on several levels of the industry, especially around how the increased actual digitization of all areas of cinematic technology has benefited the industry's concerted effort to marginalize effects artists and devalue digital labor in general.

"NOT CINEMATOGRAPHY": THE EFFECTS ARTISTS' SIDE

I have been making the argument throughout this chapter that underneath much of cinematographer's strong proclamations of what "is" and "is not" cinematography lies a concern that increasingly "going digital" potentially comes with the loss of artistic status. Redefining the cinematographer as lighting designer would mean a loss of specialization and a potential eventual lumping together with other "digital workers." As Schaefer and Cundey discussed, the role of cameras and lenses to shape light are central to cinematographers' professional and artistic identity. Again, this apprehension comes from what has happened to effects workers as they transitioned from primarily camera workers to computer screens. While cinematographers also complain interference with their image capture from workers in editorial, the digital intermediate, not to mention the director and/or the producer or star, it is the precedent of the professional downward trajectory of the effects worker that provides the most troubling pattern.[63]

Due to the prominent financial success and popular reception of early 1990s films with showstopping CGI such as *Terminator 2: Judgement Day* (Cameron, 1991) and *Jurassic Park* in 1993, the effects sector was the first begin the transition to digital technology. It would be logical to imagine that that effects' increased presence (supported by developments in digital technology) in nearly every Hollywood release would have led to the rise in professional status and appreciation for the artistry of effects workers. However, due to concerted efforts by studios to keep costs down (and also poor financial handling by the effects houses themselves), the effects industry's professional status has not enjoyed such an expected upsurge. The visual effects industry has been in well-documented turmoil over the last several years. Elsewhere,

Leon Gurevitch as well as Michael Curtin and John Vanderhoef have detailed the plight of the contemporary globalization of the "migratory" effects worker as it relates to contemporary creative labor.[64] These scholars have (rightly) put blame for the industry's woes on the larger economic forces of neo-liberal labor practices and globalization. Granting the importance of these factors, I suggest that the visual effects industry's particular history and context makes it unlike other areas of creative labor or so-called "runaway production."[65] And I argue that abetted by the convergence rhetoric of flexibility and interchangeability, it is the industry's move away from traditional camera-based cinematography onto computer work stations that has been major contributing factor to the industry's troubles.

Historically speaking, cinematographers witnessed the undercutting of effects workers' positions, who when professionally separated from the physical camera through digital technologies, lost camera-based union and guild representation. Certainly, effects workers have never enjoyed the prestige of a director of photography. However, unlike other areas of production transformed by digital technology such as editing or sound, the effects artist used to be very similar in labor category to cinematographers. As I have discussed elsewhere, there was a brief ascendency of the "effects auteur" (artists such as Douglas Trumbull and John Dykstra) in the 1970s and 1980s with the success of *Star Wars* (Lucas, 1977) and *Close Encounters of the Third Kind* (Spielberg, 1977).[66] However, effects artists have lost what little clout they were building in the 1970s and 1980s in the digital transition.

Historically from the 1910s, both groups belonged to the professional and labor category of "cameraman."[67] Initially, the on-set cameraman, along with his regular duties completed "trick effects" in the camera, ranging from irises and fades to more complex superimpositions and matte shots. Although the jobs eventually split apart through the late teens to the 1930s, what we now consider the vastly different "cinematographer" and "effects artist" categories were essentially the same job.[68] Importantly, they had the same basic apprentice-based training with the camera. Cinematographers and DPs were considered higher status in the production, but effects artists were essentially a more specialized category of cameraman.[69] The early ASC from its founding in 1919, actively recruited trick cameramen, and many, like Fred Jackman and Homer Scott served as ASC president or other officers for many of its early decades. Likewise, even well into the 1990s, prominent effects supervisors like Richard Edlund, John Dykstra, and Dennis Muren became ASC members as well as camera union members.[70] Also through the 1990s, *American Cinematographer* closely covered effects and effects artists in their feature articles. By contrast, prominent contemporary visual effects supervisors such as Roger Guyett (of ILM, VFX Supervisor *Star Wars: The Force Awakens*) Janek Sirrs (VFXS The Matrix series, *The Avengers, Batman*

Begins), John Knoll (of ILM, VFXS Star Wars prequels, The Pirates of the Caribbean series, *Pacific Rim*) and Joe Letteri (of Weta Digital, VFXS Lord of the Rings series, *Avatar*) are not ASC members.

This dramatic labor shift happened not only due to broader outside economic forces of globalization and neoliberal recategorization of creative labor. Former VES President Jeff Okun describes the historical process that split cameramen and effects artists this way:

> Special visual effects [of the 1970s. . .] was all camera people. And then with the digital revolution, or the digital tidal wave, for whatever reason they did not embrace any of this. [. . . They] were putting stuff up on monitors and filming the finished shot through a film camera that was how we got it. Then you go to Triple-I, which ultimately became Gary Demos and John Whitney, Jr.'s company, ... They used Crayex MP1 and Demos figured out a better way to get to scan an image directly onto film using a laser beam, as opposed to shooting it off a monitor and all that stuff. What you have was a divergence of who is operating the equipment. Shooting an image off a monitor is camera people versus scanning an image directly to film was more techies. [. . .] There was a split, if you will, and then on the other side of the coin the digital artists, the supervisors mostly, who weren't camera people, they were magicians. The Dykstra's are rare, that whole breed the first tier as I call them; they all came out of the camera world. The next tier was my tier, where we straddled both worlds, and then the tier after us the digital kids, they never . . . [learned to use a camera].[71]

In other words, the split happened because effects technology and personnel shifted entirely into the digital realm. Once effects artists moved away from the actual physical camera as their primary tool, they were no longer the purview of the ASC or the camera union. Moreover, the industry still recognizes as superior the cinematographer's talent with a human-finessed camera over what is misunderstood as the effects workers' "automatic" algorithm running. The major consequence of this lack of artistic recognition means today effects workers are not represented by talent agents, like cinematographers typically are, they have no labor union, their guild, the Visual Effects Society (VES), has little industry clout. Their working conditions are unregulated, forcing them into long hours of unpaid overtime and domestic disruption as they are forced to constantly displace themselves as they chase jobs across the globe, from Canada, to the UK, India, China and New Zealand for dwindling pay.[72]

For the special visual effects side, the smooth rhetoric of cooperation is doubly forced upon effects artists, first from the particular film production and then from the business at large, often competing with one another to seem easier to work with. Scratch below the surface, as several interviews I have conducted with effects workers, and a great deal of resentment regarding "credit where credit is due" is evident. And as the quotes at the beginning of the essay suggest, this resentment is rarely directed at the cinematographer, except when he doesn't credit the effects, as was the case with Claudio Miranda on *The Life of Pi*.[73] More often the anger is focused higher up, sometimes at director, but usually at the production. Much of the bad feeling comes from effects artists not being recognized as artists, but being treated like button pushers who fix problems magically with a few keystrokes. Furthermore, the characterization in the professional and fan literature as compliant service workers (as seen in the "We're just doing what the director tells us" type of rhetoric[74]) has had serious consequences for how effects companies are expected to operate.

Okun describes the pernicious effect the bland cooperation rhetoric of cooperation has on their process. Because one rarely sees frank descriptions of working relationships in the journalism or promotion, it is worth quoting at length:

> And none, to this day, none of the special extras they call them talk about the artists, or the artistry. They all talk about the computer, and all of them and *Cinefex* magazine in particular, and I love them dearly and they are good friends, but none of their stories talk about reality. All their stories talk about "we met with the director, we presented our idea, he loved it, we shot it, presented it, and we are the best buddies we go out drinking now." None of them talk about, which is when, it was on *Stargate*, when they wanted to do an article with me and this is when I became aware of it. I said "well, we are gonna tell the truth, right?" Tell the truth of how I presented 35 versions to Roland [Emmerich] of what the stargate was and he ended up firing me and I went and screwed around with a garage with the last of the resource because I had a whole day and then showed him that stuff and he goes "oh yeah that's what I want." And none of it talks about how we shot the whole trip in the stargate and then he fired me again. And that trip took a month to render. And we only had two weeks left to redo the whole thing and the tricks we pulled in order to show him essentially the same thing and let him think we were doing something new but added lens flares to it, and then he loved it. And they said "we are not interested in that story." And what they ended up writing was stuff that wasn't it. As a result of all this PR that it's all just fun and games, there is no hard work involved, it's really cool stuff, you get to hang out with really cool people, there is no trial and error, there are no failures, we bred a group of people that are like "hey, this will be fun" and so . . .[75]

In other words, the bland public rhetoric of cooperation distorts the actual problematic labor conditions, as well as obscures specific expertise. Moreover, the director and producer need to understand and respect the effects worker as an artist. If the effects team received recognition as skilled, experienced artists with similar prestige as a cinematographer, they would understand they are professionally and with talent performing an art, not providing a rote service that anyone with a computer could do. Okun calls this misperception the "Hey, my kid has a Mac" devaluation of effects artists' specialized skills and talents.

CONCLUSION: WHO PAINTS THE LIGHT?

Does it matter if the hallowed cinematographers' "painting with light" is not achieved primarily on the set with "real" light but instead done using a computer, as the *Gravity* example delineates? And does that mean that the effects worker is doing the same as the cinematographer job albeit virtually in the computer? I argue that these questions are not just local debates carried out among below-the-line workers. The specific case of the lighting designer controversy helps academics see the greater material consequences of the professional and popular rhetoric of digital convergence, and how our own rhetoric counts.

Blockbuster filmmaking's reliance on effects work should mean that cinematographers (and everyone else) fear being usurped or eclipsed by the fantasy of digital technicians wielding all-powerful digital tools.[76] However, cinematographers, who know that the digital technology used by effects workers is different from what they use, as well as that of the digital intermediate. They also know that neither effects workers nor anyone else have the ability or the professional inclination to simply digitally paint over the film with an all-encompassing digital brush. Instead, the cinematographer's concern is based in how their work is valued. They do not want to be turned into undifferentiated digital effects workers with equally poor working conditions as effects workers.[77] And to reiterate, it is not a matter of cinematographers and effects workers' skill sets actually being merged by digital technology. Instead, it is the (albeit wrong) perception that that such convergence is possible and even desirable by conglomerate Hollywood, which could lead to a similar erosion of labor protections in other areas.

The concerted conglomerate effort to rhetorically merge these disparate skill categories into "digital workers" has already had some real world consequences, most markedly with governmental subsidies meant to attract production. For example, the Australian government's subsidies program for attracting production has a significant component for what they call "PDV" (post, digital and visual effects) offsets, which includes nearly all post-production digital work, including visual effects, animation, ADR, music composition, editing, digital intermediate, and many others.[78] Although cinematography is not included in this list, it does encompass much of what would be considered the "lighting designer's purview."

Beyond what might be considered internecine squabbling, highlighting this dispute has important consequences to destabilizing the smooth and systematized Hollywood business rhetoric of digital convergence as a subset of conglomeration and synergy. In other words, the studio forces the workers into rhetoric of convergence as digital interchangeability, but actual jobs require non-transposable specialization. I argue this corporate discourse, which emerges as bland claims of cooperation, demonstrates that convergence is not a technological condition of the digital as it is usually characterized. And further, by understanding digital convergence as a rhetoric, we understand there is no such thing as "the digital," but many digitals with incompatible operating systems, pipelines, and software that just as frequently serve to separate teams of workers, rather than foster cooperation or solidarity. The positivized public corporate rhetoric of vertical integration is directed at consumers to create the illusion of convenience, speed and access associated with digital delivery systems, but also permeates the rhetoric of cinema professionals, forced to "play along" in an increasingly precarious and unstable workplace. More specifically, I argue that professional discourse on the relation between cinematography and special visual effects is not simply a mark of the forces of globalization and the neoliberal economy. It is a site of a unique and long history of interrelation, whose history points to a downward slip into labor precarity and lack of status. In it, cinematographers see what effects artists have lost, and the professional stakes are considerable.

NOTES

1. In addition to thanking the volume editors Lindsay Coleman, Daisuke Miyao and Roberto Schaefer for their comments and suggestions, I would like to especially thank Roberto Schaefer, ASC, AIC, Dean Cundey, ASC, and Jeffery Okun for allowing me to interview them for this essay. I would also like to thank the University of Illinois research board for providing research funds that made this essay possible.

2. I say "credited DP" because Fiore reportedly only shot the live action actors sequences, while Cameron collaborator Vince Pace, credited as "Director of Photography Los Angeles," according to Schaefer, "shot probably 80 percent of the movie, [. . .] shot with Cameron for months before Mauro even came on the film.

3. Adding, "I mean, I admire Chivo, and he's an amazing cinematographer and he does great work." Roberto Schaefer, personal telephone interview with the author, July 22, 2015.

4. Benjamin B, "Facing the Void: *Gravity,*" *American Cinematographer*, November 2013, 36–37.

5. The film production pipeline, following the traditional studio chronology, is loosely defined as pre-production, production, post-production, but also encompasses how and in what order design, research and development, then shooting and then digitally produced material and manipulation is handled throughout the production.

6. As a way to rhetorically bolster the labor position of effects artists, I have proposed the category "special visual effects" and designation "special visual effects artist." I am aware it is not in line with industry designations. Within the industry, the term special effects or SFX historically refers to physical effects, like pyrotechnics, fire, snow, gunshot squips. Post-production effects work has been termed Visual Effects, or VFX. However, I propose special visual effects artist as an historically grounded term that honors SVFX artists as specialized labor as well as skilled artists. Julie Turnock, "The Special Effects Business is an Oxymoron": Special Visual Effects, Blockbuster Filmmaking, and the Problem of Independence" paper presented at the annual meeting for the Society for Cinema and Media Studies, Montreal, March 25–29, 2015.

7. Schatz defines "conglomerate Hollywood" as the hegemonic control since the mid-1980s of the giant transnational media companies (then, General Electric, Disney, News Corp, Time Warner, Sony and Viacom). These companies not only control the vast majority of film production, but also other media forms such as television, music, print publishing, video and computer games, consumer electronics and theme parks. Thomas Schatz. "New Hollywood, New Millenium." In Warren Buckland (ed) *Film Theory and Contemporary Hollywood Movies* (Routledge 2009) 21.

8. Benjamin B, "Facing the Void," 36–37.

9. Sam Gaskin, "Christopher Doyle Interview Part 2: "Life of Pi" Oscar is an Insult to Cinematography," last modified March 5, 2013, http://sea.blouinartinfo.com/news/story/874483/christopher-doyle-interview-part-2-life-of-pi-oscar-is-an

10. Bruce Branit, Facebook post, February 25, 2013, https://www.facebook.com/bruce.branit/posts/10200665318986444

11. John Alton, ASC, influentially published articles and books on cinematography, characterizing his work as "Painting with Light." Reprinted recently as John Alton, *Painting with Light* (Berkeley, CA: University of California Press, 1995). Vittorio Storaro, ASC, AIC, has frequently referred to his cinematography as "writing with light." *Scrivere con la luce/Writing with Light*. New York: Aperture, 2003.

12. Donald Crafton. *The Talkies: American Cinema's Transition to Sound, 1926–1931*. Vol. 4. University of California Press, 1999: 23.

13. Henry Jenkins and Mark Deuze, "Convergence Culture." *Convergence-London*- 14.1, 2008, 5. Keeping to industrial characterizations, I will not be concerned with the most typical definitions of convergence as Henry Jenkins has influentially depicted it, with digital technologies allowing greater erasing of the boundaries between "professional" and "non-professional" users of media, with an emphasis on nonprofessional users of media's ability to "participate" in broader media forms by reappropriation. In the case of effects work, Jenkins-style participation is operative, for example, the effects industry boogeyman of movie fan-enthusiast being able to participate in the effects business at a very basic level with a laptop and appropriate software, but to the advantage of conglomerate Hollywood. For the effects industry, this nonprofessionalization has led to the "I'd do it for free" phenomenon among entry level works, meaning a glut of non-unionized, cut rate effects workers that the larger firms have to compete with. While this may be seen as an example of "free market" ingenuity, the "one guy with a laptop" is also working against his own interests.

Moreover, the Bolter and Grusin style of convergence, with an emphasis on digital technology's ability to more easily appropriate and transform "analog" material. David J. Bolter and Richard Grusin, *Remediation: Understanding New Media* (Cambridge, MA: MIT Press, 2000).

Finally, Derek Johnson has also discussed comics giant Marvel's corporate strategies in terms of convergence of story worlds, or "collisions of extracinematic sectors of all types." Derek Johnson, "Cinematic Destiny: Marvel Studios and the Trade Stories of Industrial Convergence." *Cinema Journal* 52, no. 1, 2012: 2.

14. Stephen Prince, *Digital Visual Effects in Cinema: The Seduction of Reality*. Rutgers University Press, 2011.

15. For more on the issue of international digital migrant workers, see Leon Gurevitch, "The Straw That Broke the Tiger's Back?: Skilled Labor, Social Networks and Protest in Digital Workshops of the World." *The Routledge Companion to Labor and Media*, Routledge: 2015, 190–201.

16. David Bordwell, Janet Staiger, and Kristin Thompson. *The Classical Hollywood Cinema*. New York: Columbia University Press, 1985; Gomery, Douglas. *The Hollywood Studio System: a History*. British Film Institute, 2005; Tino Balio, *Grand Design: Hollywood as a Modern Business Enterprise, 1930–1939*. Vol. 5. University of California Press, 1995.

17. More recent historical work on the studio system has complicated the notion of the Fordist model, for example Brian Johnson. *Studios Before the System: Architecture, Technology, and the Emergence of Cinematic Space*. Columbia University Press, 2015; and Emily Carman. *Independent Stardom: Freelance Women in the Hollywood Studio System*. University of Texas Press, 2015.

18. Balio, *Grand Design*; "The Dream Factory" as a term was popularized by Powdermaker, Hortense. *Hollywood, the dream factory: An anthropologist looks at the movie-makers*. Boston: Little, Brown, 1950.

19. Crafton, *The Talkies*: 182.

20. Along with Carman, Danae Clark. *Negotiating Hollywood: The Cultural Politics of Actors' Labor*. University of Minnesota Press, 1995; Gerald Horne, *Class Struggle in Hollywood, 1930–1950: Moguls, Mobsters, Stars, Reds, and Trade Unionists*. University of Texas Press, 2013; Murray Ross, "Labor relations in Hollywood." *The ANNALS of the American Academy of Political and Social Science* (1947): 58–64; Matt Stahl, "Cultural labor's "democratic deficits": Employment, autonomy and alienation in US film animation." *Journal for Cultural research* 14, no. 3 (2010): 271–293.

21. As Andrew Ross, *Nice Work If You Can Get It: Life and Labor in Precarious Times*. NYU Press, 2009; Richard Florida, *The Rise of the Creative Class—Revisited: Revised and Expanded*. Basic books, 2014.

22. Lev Manovich. *The Language of New Media*. MIT press, 2001.

23 Marcus Hawkins, "What is a Raw File? We Explain the Pros and Cons," Digital Camera World, last updated June 8, 2015, http://www.digitalcameraworld.com/2015/06/08/what-is-a-raw-file-we-explain-the-pros-and-cons/

24. Schaefer, personal interview with the author.

25. John Thornton Caldwell, *Production Culture: Industrial Reflexivity and Critical Practice in Film and Television* (Durham: Duke University Press, 2008) 195.

26. Caldwell, *Production Culture.* 5.

27. In a discussion of adapting to changing exhibition platforms, Crudo frames this tough love as beneficial to cinematographers: Intent is all that matters. Some may find this frightening, but it actually strengthens the cinematographer's position. . . . [our creative canvases] are not united by technology—they're united by our creative spark." Richard Crudo, ASC, "President's Desk" *AC*, April 2016: 12.

28. Patrick Keating discusses the early efforts of cinematographers (from the silent era to the 1940s) to be recognized as artists, not just technicians. Patrick Keating. *Hollywood Lighting from the Silent Era to Film Noir.* Columbia University Press, 2009: 15, 257–264.

29. Schaefer, interview with the author

30. "What I am up in arms about is the academy and worse, the ASC, among others giving the credit and awards to the wrong people. The VFX and animators should be awarded, not the DP." Schaefer, interview with the author.

31. Benjamin B, "Facing the Void," 36–37.

32. As Schaefer put it, such a situation "is rarer than a blue moon on Friday the thirteenth." Schaefer, interview with the author.

33. Cinematographers a typically not paid for post-production work, As Jeff Cronenweth, ASC, noted in 2013, in regard to his work on the film *Hitchcock* (Gervasi, 2012):

> With photochemical timing, you'd donate your time when you were called into the lab for a screening with the color timer, you'd give notes as the film was played in its entirely, the timer would apply those notes, and then, a couple of days later. . . . But in todays world, you're sitting in the color suite 12 hours a day, coloring every frame. [Production] usually books the room for 2 weeks or so, so you basically have to take two weeks off from any other jobs to finish the movie. It requires full days and full weeks, and it's a real liability to do it gratis. Luckily, I was paid for this work on *Hitchcock*, but that's not the norm in the industry, and that has to change. It's up to the cinematographer to protect the image and get the best image possible for the director and producer, but that's becoming a much bigger job than it used to be.

Jay Holben "An Auteur's Angst" *AC* Jan 2013: 88

34. David S. Cohen, "Blockbusters Take Toll On F/X Shops," *Variety* Posted: Fri., May 25, 2007.

35. Cohen, "Blockbusters Take Toll"

36. There is a long history of professional journals such as *American Cinematographer* floating new terminology for job categories, see Turnock, "Patient research on the Slapstick Lots: From Trick Men to Special Effects Artists in Silent Hollywood," *Early Popular Visual Culture* 13:2 (2015): 152–173.

37. Benjamin B, "Facing the Void," 36–37.

38. For recent academic work on the index in the digital age, see Manovich, David N, Rodowick, *The Virtual Life of Film.* Harvard University Press, 2009; and Laura Mulvey. *Death 24x a Second: Stillness and the Moving Image.* Reaktion Books, 2006.

39. Of course, Bazin's position on the index is complex. See Dan Morgan for a discussion of Bazin. Daniel Morgan, "Rethinking Bazin: Ontology and Realist Aesthetics." *Critical Inquiry* 32, no. 3 (2006): 443–481; and Francesco Casett,. "Sutured reality: Film, from photographic to digital." (2011).

40. Many cinematographers, even those who revere shooting on film, have decided the negatives outweigh the positive. For example, in interviews for *Hail Caeser!*" Roger Deakins frequently declared (somewhat regretfully) "Sorry, film is over." http://www.indiewire.com/2016/01/film-im-sorry-its-over-roger-deakins-talks-hail-caesar-working-with-coens-and-going-digital-86166/; and http://variety.com/2016/artisans/in-contention/hail-caesar-roger-deakins-celluloid-1201687528/.

41. Tom Gunning has discussed this along academic terms in Gunning, "Moving away from the index: Cinema and the impression of reality." *differences* 18, no. 1 (2007): 29–52.

42. Roberto Schaefer, personal interview with the author. See the *American Cinematographer* podcast interview with Claudio Miranda, for his characterization of the limited interaction or collaboration between the effects team and DP. "Life of Pi podcast" circa Nov 2012, https://www.theasc.com/site/podcasts/life-of-pi-claudio-miranda-asc/

43. Schaefer, personal interview with the author.

44. See interviews in the documentaries *Visions of Light* (Glassman and McCarthy, 1992) *Side by Side* (Kenneally, 2012) for examples of this rhetoric.

45. Dean Cundey, personal interview with the author, Pasadena, CA, August 17, 2105.

46. Cundey, personal interview with the author.

47. Cundey, personal interview with the author.

48. Cundey, personal interview with the author.

49. For how Calahan characterizes her work, see Jon Silberg, "The Good Dinosaur: Q&A with Sharon Calahan" *American Cinematographer* (online exclusive) December 2015, http://www.theasc.com/ac_magazine/December2015/TheGoodDinosaur/page1.php

50. Cundey, personal interview with the author.

51. Cundey, personal interview with the author. For another voice who agrees with Cundey, in the "Close up" feature with Greig Fraser, ASC, ACS [cinematographer for *Rogue One*] in response to the question, "What recent books films artwork have inspired you?" he answered: ". . . *Inside out*.. [I was] blown away by the quality of the cinematography—this has not received nearly enough attention in this category." *AC* Feb 2016: 96.

52. Schaefer, personal interview with the author

53. Schaefer, personal interview with the author

54. For one of many examples, see Jon D. Witmer, "All Together Now," on Marvel's the Avengers, *AC,* June 2012,35.

55. Cite cinefex, amcin

56. Benjamin B, "Facing the Void," 38.

57. Benjamin B, "Facing the Void," 38.

58. Benjamin B, "Facing the Void," 41,43.

59. Benjamin B, "Facing the Void," 43.

60. Benjamin B, "Facing the Void," 44.

61. Benjamin B, "Facing the Void," 48.

62. Benjamin B, "Facing the Void," 48–49.

63. For an account characterizing DPs' fears of their eroding power and of new technology and competition between DPs and DI colorists, see Jean-Pierre Geuens, "Angels of Light," Carol Vernallis, Amy Herzog, and John Richardson, eds. *The Oxford Handbook of Sound and Image in Digital Media* (Oxford: Oxford University Press, 2013), 45–60, especially 50.

64. Curtin and Vanderhoef, "A Vanishing Piece of the Pi: The Globalization of Visual Effects Labor," *Television and New Media* Vol 16 (3) 219–239; and Gurevitch, "Straw that Broke . . ."

65. As Toby Miller et al. have described it. Toby Miller, Nitin Govil, John McMurria, Richard Maxwell, and Ting Wang. *Global Hollywood 2*. BFI Publishing, 2005.

66. See Turnock, "Special Effects: Auteur Renaissance, 1968–1980," in *Editing and Special Visual Effects*: *Behind the Silver Screen*, volume eds. Charles Keil and Kristen Whissel, series ed. Jon Lewis (Rutgers University Press/Academy of Motion Picture Arts and Sciences, 2016): 116–128.

67. And to be clear, those operating the camera since the 1910s were, and continue to be, predominantly men.

68. See Turnock, "Patient Research." Camera unions did not represent all effects workers, of course. Matte painters typically belonged to the art director's guild and were members of Art Director's union.

69. See ASC member lists, for example *AC* Feb 1922, 56; *AC* April 1924: 'Membership of the ASC to Date' 27; *AC* Oct 1927, 14.

70. And many of them, including Edlund, consider themselves more cinematographer than effects artist.

71. Okun, personal interview with the author.

72. While cinematographers have a high position in the production hierarchy they have worked hard to attain and are working hard to maintain, the protesting effects workers are speaking from a historically diminished vantage point. Effects workers understand, like cinematographers, that the currency of Hollywood is perceived as a unique artistic specialization. They have been very active in trying to raise their position rhetorically and through organized labor demonstrations. Their most prominent attempt to make themselves visible has come in labor action spurred by protests at the 2013 Academy awards over the *Life of Pi* Oscar for visual effects, as well as the picketing of the award ceremony in 2014, providing them with their most publically visible form to date. Both actual and virtual picket lines formed around the special effects world, including Wellington, Los Angeles, Vancouver, London and elsewhere, and blogs and twitter worked to organize these geographically diverse groups. In addition to unionization into a real guild, many advocate for the end of international subsidies, or representation by agents like cinematographers have. However, it is hard enough to collectivize the LA-based effects workers; international labor law and tax subsidies make collectivizing especially difficult in the far flung SVFX community. As I have discussed elsewhere, their work continues. As in Julie Turnock, "The Special Effects Business is an Oxymoron;" and Gurevitch, "Straw that Broke. . ."

73. Generally speaking, the cinematographers I spoke to and also in the professional publications such as *American Cinematographer* report positive relationships with effects supervisors.

74. As Robert Richardson says of his work on *Django Unchained* (Tarantino, 2012, despite, or because of, the rumored discord with the director) "The director is the one with the vision, and we serve him or her. The only path to creating a great film is to support the director." Iain Stasukevich, "Once Upon a Time in the West" *AC*, Jan 2013: 34.

75. Okun, personal interview with the author.

76. Many articles of the late 1990s and early 2000s somewhat prematurely reflected this notion, especially around the digital intermediate, for example Prince, Stephen. "The Emergence of Filmic Artifacts: Cinema and Cinematography in the Digital Era." *Film Quarterly* 57, no. 3 (2004): 24–33"; John Belton, "Painting by the Numbers: The Digital Intermediate." *Film Quarterly* 61, no. 3 (2008): 58–65; Aylish Wood. "Pixel visions: Digital intermediates and micromanipulations of the image." *Film Criticism* 32, no. 1 (2007): 72.

77. Even ASC president Richard Crudo writes of the problem of long hours and "draconian" working conditions facing contemporary cinematographers, blaming it on the ASC's excellence: "We've become so good at our jobs that we've made the delivery of first-class production look easy no matter how onerous the circumstances." Richard Crudo, "President's Desk" *AC* March 2016: 12.

78. See the Australian government white paper on PDV offsets, especially pages 16–17. http://arts.gov.au/sites/default/files/about/Guidelines%20location%20pdv%20offsets%20December%202013.pdf; http://www.ausfilm.com.au/why-film-in-australia/financial-incentives/post-digital-and-visual-effects-production-pdv-offset/

BIBLIOGRAPHY

Alton, John. *Painting with Light.* Berkeley, CA: University of California Press, 1995.

Australian Government, Attorney General's Department, Ministry for the Arts "Guidelines to Australian Screen Production Incentive," 2013 http://arts.gov.au/sites/default/files/about/Guidelines%20location%20pdv%20offsets%20December%202013.pdf

B[sic], Benjamin. "Facing the Void: *Gravity,*" *American Cinematographer*, November 2013, 36–49.

Balio, Tino. *Grand Design: Hollywood as a Modern Business Enterprise, 1930–1939.* Vol. 5. University of California Press, 1995.

Belton, John. "Painting by the Numbers: The Digital Intermediate." *Film Quarterly* 61. 3 (2008): 58–65.

Bolter, David J. and Richard Grusin. *Remediation: Understanding New Media.* Cambridge, MA: MIT Press, 2000.

Bordwell, David, Janet Staiger, and Kristin Thompson. *The Classical Hollywood Cinema.* New York: Columbia University Press, 1985.

Branit, Bruce. Facebook post, February 25, 2013. https://www.facebook.com/bruce.branit/posts/10200665318986444

Caldwell, John Thornton. *Production Culture: Industrial Reflexivity and Critical Practice in Film and Television.* Durham: Duke University Press, 2008.

Carman, Emily. *Independent Stardom: Freelance Women in the Hollywood Studio System.* Austin, TX: University of Texas Press, 2015.

Casetti, Francesco. "Sutured Reality: Film, from Photographic to Digital." *October* Fall (2011): 95–106.

Clark, Danae. *Negotiating Hollywood: The Cultural Politics of Actors' Labor*. Minneapolis: University of Minnesota Press, 1995.

Cohen, David S. "Blockbusters Take Toll On F/X Shops," *Variety* Posted: Fri., May 25, 2007.

Crafton, Donald. *The Talkies: American Cinema's Transition to Sound, 1926–1931*. Vol. 4. Berkeley, CA: University of California Press, 1999.

Crudo, Richard. ASC, "President's Desk." *AC*, March 2016: 12.

———. "President's Desk." *AC* April 2016: 12

Cundey, Dean. Personal interview with the author, Pasadena, CA, August 17, 2105.

Curtin, Michael and John Vanderhoef. "A Vanishing Piece of the Pi: The Globalization of Visual Effects Labor." *Television and New Media* 16.3 (2014): 219–239.

Florida, Richard. *The Rise of the Creative Class--Revisited: Revised and Expanded.* Philadelphia: Basic Books, 2014.

Fraser, Greig. ASC, ACS "Close Up," *AC*, Feb 2016: 96.

Gaskin, Sam. "Christopher Doyle Interview Part 2: "Life of Pi" Oscar is an Insult to Cinematography," last modified March 05, 2013. http://sea.blouinartinfo.com/news/story/874483/christopher-doyle-interview-part-2-life-of-pi-oscar-is-an

Gomery, Douglas. *The Hollywood Studio System: A History*. London: BFI, 2005.

Geuens, Jean-Pierre. "Angels of Light." In Carol Vernallis, Amy Herzog, and John Richardson, eds. *The Oxford Handbook of Sound and Image in Digital Media.* Oxford: Oxford University Press (2013): 45–60.

Gunning, Tom. "Moving Away from the Index: Cinema and the Impression of Reality." *differences* 18. 1 (2007): 29–52.

Gurevitch, Leon. "The Straw That Broke the Tiger's Back?: Skilled Labor, Social Networks and Protest in Digital Workshops of the World." in ed. Richard Maxwell. *The Routledge Companion to Labor and Media.* London: Routledge (2015): 190–201.

Hawkins, Marcus. "What is a Raw File? We Explain the Pros and Cons," Digital Camera World, last updated June 8, 2015, http://www.digitalcameraworld.com/2015/06/08/what-is-a-raw-file-we-explain-the-pros-and-cons/

Hemphill, Jim. "Podcast: *Life of Pi*" Episode 42, c. Nov 2012. https://www.theasc.com/site/podcasts/life-of-pi-claudio-miranda-asc/

Holben, Jay. "An Auteur's Angst" *AC* Jan 2013: 78–88.

Horne, Gerald. *Class Struggle in Hollywood, 1930–1950: Moguls, Mobsters, Stars, Reds, and Trade Unionists*. Austin, TX: University of Texas Press, 2013.

Jacobson, Brian. *Studios Before the System: Architecture, Technology, and the Emergence of Cinematic Space*. New York: Columbia University Press, 2015.

Jenkins, Henry and Mark Deuze. "Convergence Culture." *CONVERGENCE-LONDON-* 14.1 (2008): 5–12.

Jewell, Richard B. "How Howard Hawks Brought Baby Up: An Apologia for the Studio System." *Journal of Popular Film and Television* 11.4 (1984): 158–165.

Johnson, Derek. "Cinematic Destiny: Marvel Studios and the Trade Stories of Industrial Convergence." *Cinema Journal* 52.1 (2012): 1–24.

Keating, Patrick. *Hollywood Lighting from the Silent Era to Film Noir*. New York: Columbia University Press, 2009.

Manovich, Lev. *The Language of New Media*. Cambridge, MA: MIT press, 2001.
Miller, Toby, Nitin Govil, John McMurria, Richard Maxwell, and Ting Wang. *Global Hollywood 2*. London: BFI Publishing, 2005.
Morgan, Daniel. "Rethinking Bazin: Ontology and Realist Aesthetics." *Critical Inquiry* 32. 3 (2006): 443–481.
Mulvey, Laura. *Death 24x a Second: Stillness and the Moving Image*. London: Reaktion Books, 2006.
Okun, Jeffery. Personal interview with the author. August 8, 2105, Encino, CA.
Powdermaker, Hortense. *Hollywood, the Dream Factory: An Anthropologist Looks at the Movie-Makers*. Boston: Little, Brown, 1950.
Prince, Stephen. *Digital Visual Effects in Cinema: The Seduction of Reality*. New Brunswick, NJ: Rutgers University Press, 2011.
———. "The Emergence of Filmic Artifacts: Cinema and Cinematography in the Digital Era." *Film Quarterly* 57. 3 (2004): 24–33.
Rodowick, David N. *The Virtual Life of Film*. Harvard University Press, 2009.
Ross, Andrew. *Nice Work If You Can Get It: Life and Labor in Precarious Times*. New York: NYU Press, 2009.
Ross, Murray. "Labor Relations in Hollywood." *The ANNALS of the American Academy of Political and Social Science* (1947): 58–64.
Schaefer, Roberto. Personal telephone interview with the author, July 22, 2015.
Schatz, Thomas. "New Hollywood, New Millenium." In Warren Buckland, ed. *Film Theory and Contemporary Hollywood Movies* (Routledge 2009) 21.
Silberg, Jon. "The Good Dinosaur: Q&A with Sharon Calahan" *American Cinematographer* (online exclusive) December 2015. http://www.theasc.com/ac_magazine/December2015/TheGoodDinosaur/page1.php
Stahl, Matt. "Cultural Labor's "Democratic Deficits": Employment, Autonomy and Alienation in US Film Animation." *Journal for Cultural Research* 14. 3 (2010): 271–293.
Stasukevich, Iain. "Once Upon a Time in the South" *AC*, Jan 2013, 32–49.
Storaro, Vittorio, ASC, AIC. *Scrivere Con la Luce/Writing with Light*. New York: Aperture, 2003.
Turnock, Julie. "Patient Research on the Slapstick Lots: From Trick Men to Special Effects Artists in Silent Hollywood." *Early Popular Visual Culture,* 13. 2 (2015): 152–173.
———."Special Effects: Auteur Renaissance, 1968–1980," In eds. Charles Keil and Kristen Whissel. *Editing and Special Visual Effects: Behind the Silver Screen*. Rutgers University Press/Academy of Motion Picture Arts and Sciences (2016): 116–128.
———."The Special Effects Business is an Oxymoron": Special Visual Effects, Blockbuster Filmmaking, and the Problem of Independence." unpublished paper presented at the annual meeting for the Society for Cinema and Media Studies, Montreal, March 25–29, 2015.
Witmer, Jon D. "All Together Now," *AC,* June 2012, 34–51.
Wood, Aylish. "Pixel Visions: Digital Intermediates and Micromanipulations of the Image." *Film Criticism* 32. 1 (2007): 72–94.

Index

About the Editors and Contributors

Lindsay Coleman is a writer and academic based in Melbourne, Australia. His published anthologies include *The Philosophy of Pornography* (Scarecrow Press) and *Sex and Storytelling in Modern Cinema* (SUNY Press). Upcoming books include anthologies on film composition, the practice of cinematography, and the filmmaker Shohei Imamura. Lindsay is also a journalist for the Australian Cinematographer magazine, and is in the process of writing a television series on the history of visual effects in cinema. Lindsay is also a senior instructor for *Manhattan Review*.

Ceri Higgins has worked in film and television production in the UK, Mexico and the United States for over thirty-five years. As vice president of Telenovelas Historicas at Televisa she oversaw the production of two major international drama series, *El vuelo del Aguila* and *La antorcha encendida.* As an author and academic her writing includes her book on the cinematographer Gabriel Figueroa *'Nuevas perspectivas',* articles and chapters. She is currently course leader of Film and Television at the UAL London.

Patrick Keating is associate professor in the Department of Communication at Trinity University, where he teaches courses in film studies and video production. He is the author of *Hollywood Lighting from the Silent Era to Film Noir* and the editor of *Cinematography*. For his ongoing research on the history of camera movement, he has been awarded an Academy Film Scholars grant and a Frederick Burkhardt Fellowship for Recently Tenured Scholars.

Evan Lieberman is associate professor of Film and Video at Cleveland State University where he teaches courses in filmmaking, cinematography, and film theory. He received his MFA in Screenwriting from the American

Film Institute's Center for Advanced Film and Television Studies and his PhD in Film Studies/American Studies from Emory University. In addition to teaching and writing on an array of subjects ranging from the television situation comedy and cinematography to Mexican wrestling movies, he maintains an active production career as a writer, producer, director, and cinematographer, having won dozens of awards for his film work at festivals around the world. He is currently working on a book titled *The Art of the Moving Image* with David Cook and his most recent film *Sorry I Wasn't There Last Night*, which he cowrote, coproduced and served on as director of photography is in post-production.

Daisuke Miyao is professor and the Hajime Mori Chair in Japanese Language and Literature at the Department of Literature, the University of California, San Diego. He is the author of *The Aesthetics of Shadow: Lighting and Japanese Cinema* (Durham, NC: Duke University Press, 2013), *Eiga wa neko dearu: Hajimeteno cinema sutadizu* (Cinema is a cat: An introduction to cinema studies) (Tokyo: Heibon sha, 2011), and *Sessue Hayakawa: Silent Cinema and Transnational Stardom* (Durham, NC: Duke University Press, 2007). Miyao is also the editor of *The Oxford Handbook of Japanese Cinema* (Oxford: Oxford University Press, 2014). He is currently writing a book, which is tentatively titled *Japonisme and the Birth of Cinema*.

Jakob Isak Nielsen was born in 1975. He is associate professor and head of the department at Media and Journalism Studies, School of Communication and Culture, at Aarhus University (AU) in Denmark. He is co-editor of and contributor to *Fjernsyn for viderekomne – de nye amerikanske tv-serier* (Turbine 2011) along with being the founding editor of *16:9* since 2003 (www.16-9.dk). Besides publications in *16:9*, his work has been published in *Northern Lights: Film & Media Studies Yearbook* (Intellect 2016), *A Companion to Fritz Lang* (Wiley-Blackwell 2015), various anthologies and journals such as *The Journal of Popular Television, Short Film Studies, Kosmorama, P.O.V., MedieKultur and Passage.* He is also a member of different administrative committees: The Forum for Media Studies (DK), The Museum Council at the Danish Film Institute, Copenhagen, and the Educational Council for Multiplatform Storytelling and Production, VIA University College.

Peter H. Rist, PhD, is professor of Film Studies in the Mel Hoppenheim School of Cinema at Concordia University in Montréal, Québec, Canada, where he was the chair (head) for eight years. He has recently written a chapter for a book on the *Poetics of Chinese Cinema,* and his *Historical Dictionary of South American Cinema* was published in 2014 by Rowman and Littlefield. He has edited two other books (one on Canadian Cinema) and has had book chapters published on sub-Saharan African, Chinese, Taiwanese, Hong Kong, and Korean films/filmmakers, and has also written a chapter titled "The Presence (and Absence) of Landscape in Silent East-Asian Films" in the book *Landscape and Film*. He was a member of the FIPRESCI juries at the 2014 Hong Kong International Film Festival, and the 2015 World Film Festival, Montreal. He reports regularly on film festivals and conducts interviews for www.offscreen.com.

Carlos Rojas is professor of Chinese cultural studies; Gender, Sexuality, and Feminist Studies; and Arts of the Moving Image at Duke University. He is the author of three books, including *Homesickness: Culture, Contagion, and National Transformation*, and is also the coeditor of five books, including, with Eileen Cheng-yin Chow, *The Oxford Handbook of Chinese Cinemas*.

Roberto Schaefer, ASC/AIC, attended art school and received a BFA degree in multimedia and conceptual art with a minor in photography. He has been making moving images since childhood using his father's 8mm camera. These included early music videos, dramatic and experimental shorts. After a four-year hiatus traveling the world and discovering life Roberto started as a producer of TV commercials in New York. At the same time he took side jobs shooting feature news documentaries for European TV outlets and short films for friends and filmmakers in the city. In 1982 Roberto moved to Rome, Italy, to shoot a feature film for a friend. In 1992, after 3 features, hundreds of TV commercials, music videos and documentaries while based in Roma and Milano, he moved to Los Angeles to try the Hollywood film world. Since then he has shot more than a dozen films, many hundreds of music videos and TV commercials, documentaries and more. Roberto has always continued to do his still photography and art. He was nominated for a CableAce award for Best Cinematography for the 1994 Showtime movie of the week "Roadracers." In 2000 the AFI honored his work on "Best in Show" as one of the top ten Movies of the Year. He was nominated for a BAFTA award for Best Cinematography for the 2003 film "Finding Neverland." Roberto has continued his work in the motion picture industry on many independent and studio pictures, including "Stranger Than Fiction," "The Kite Runner," "Quantum of Solace," "Leaves of Grass" and "The Host." In 2013 he received the Golden Frog for Cinematographer-Director duo Award with Marc Forster for their nine film, fifteen-year collaboration. Roberto also collaborated for the umpteenth time with Christopher Guest on the HBO series "Family Tree." In 2014 he went on to photograph "Term Life," "Miles Ahead" and the upcoming "Geostorm."

Julie Turnock is associate professor of Media and Cinema Studies at the University of Illinois, Urbana/Champaign, in the College of Media. She is the author of *Plastic Reality: Special Effects, Technology, and the Emergence of 1970s Blockbuster Aesthetics,* from Columbia University Press, and has published on special effects, spectacle and technology of the silent and studio era, the 1970s, and recent digital cinema in *Cinema Journal, Film History, Film Criticism,* and *New Review of Film and Television Studies.*

www.ingramcontent.com/pod-product-compliance
Ingram Content Group UK Ltd.
Pitfield, Milton Keynes, MK11 3LW, UK
UKHW020212290926
15470UKWH00001B/146

9 781498 524278